T0228082

REMEMBERING, FORGETTING
AND CITY BUILD

Re-materialising Cultural Geography

Dr Mark Boyle, Department of Geography, University of Strathclyde, UK,
Professor Donald Mitchell, Maxwell School, Syracuse University, USA and
Dr David Pinder, Queen Mary University of London, UK

Nearly 25 years have elapsed since Peter Jackson's seminal call to integrate cultural geography back into the heart of social geography. During this time, a wealth of research has been published which has improved our understanding of how culture both plays a part in, and in turn, is shaped by social relations based on class, gender, race, ethnicity, nationality, disability, age, sexuality and so on. In spite of the achievements of this mountain of scholarship, the task of grounding culture in its proper social contexts remains in its infancy. This series therefore seeks to promote the continued significance of exploring the dialectical relations which exist between culture, social relations and space and place. Its overall aim is to make a contribution to the consolidation, development and promotion of the ongoing project of re-materialising cultural geography.

Also in the series

Doing Family Photography
The Domestic, The Public and The Politics of Sentiment
Gillian Rose
ISBN 978 0 7546 7732 1

Cultural Capitals
Revaluing The Arts, Remaking Urban Spaces
Louise C. Johnson
ISBN 978 0 7546 4977 9

Critical Toponymies
The Contested Politics of Place Naming
Edited by Lawrence D. Berg and Jani Vuolteenaho
ISBN 978 0 7546 7453 5

Cultural Landscapes of Post-Socialist Cities
Representation of Powers and Needs
Mariusz Czepczynski
ISBN 978 0 7546 7022 3

Towards Safe City Centres?
Remaking the Spaces of an Old-Industrial City
Gesa Helms
ISBN 978 0 7546 4804 8

Remembering, Forgetting and City Builders

Edited by

TOVI FENSTER
Tel Aviv University, Israel

and

HAIM YACOBI
Ben Gurion University, Israel

Routledge
Taylor & Francis Group

LONDON AND NEW YORK

First published 2010 by Ashgate Publishing

Published 2016 by Routledge
2 Park Square, Milton Park, Abingdon, Oxfordshire OX14 4RN
711 Third Avenue, New York, NY 10017, USA

First issued in paperback 2016

Routledge is an imprint of the Taylor & Francis Group, an informa business

British Library Cataloguing in Publication Data
Remembering, forgetting and city builders. --
  (Re-materialising cultural geography)
  1. Collective memory and city planning--Case studies.
  2. Public spaces--Social aspects--Case studies.
  I. Series II. Fenster, Tovi. III. Yacobi, Haim.
  711.1'3-dc22

Library of Congress Cataloging-in-Publication Data
Fenster, Tovi.
  Remembering, forgetting and city builders / by Tovi Fenster and Haim Yacobi.
    p. cm.
  Includes index.
  ISBN 978-1-4094-0667-9 (hbk) -- ISBN 978-1-4094-0688-4 (ebk)
  1. Human geography--Case studies. 2. Urban geography--Case studies. 3. City planning--Case studies. 4. Geographical perception--Case studies. 5. Collective memory--Social aspects--Case studies. I. Yacobi, Haim. II. Title.
  GF50.F46 2010
  304.2'3--dc22

                                                                    2010010766

ISBN 13: 978-1-138-27868-4 (pbk)
ISBN 13: 978-1-4094-0667-9 (hbk)

Contents

List of Figures

List of Contributors

Safa Abu-Rabia is a Palestinian-Bedouin citizen of Israel, an anthropologist and doctoral student at the Ben Gurion University of the Negev, in the Department of Interdisciplinary Studies. She is currently completing a study of the construction of historical discourse by Bedouin-Arab women of the 1948 generation in the Negev. Her research interests include the construction of identity and memory among Bedouin Arabs in the Negev in light of the Nakba (1948 Palestinian catastrophe), Arab feminism, and race and racism within the context of gender in gendered Bedouin-Arab society.

Efrat Eizenberg received her PhD in Environmental Psychology from the City University of New York. She is a post-doctoral researcher at the Department of Geography and Human Environment at Tel Aviv University. Her interests are the politics of space, collective action and resistance, and environmental justice.

Tovi Fenster is the Head of the Planning for the Environment with Communities (PEC) Lab at the Department of Geography and Human Environment, Tel Aviv University. She has been the Head of NCJW Women and Gender Studies Programme (2007–2009). She publishes articles and book chapters on ethnicity, citizenship and gender in planning and development. She is the editor of *Gender, Planning and Human Rights* (1999, Routledge) and the author of *The Global City and the Holy City: Narratives on Knowledge, Planning and Diversity* (2004, Pearson). She is one of the founders and the first Chair of Bimkom – Planners for Planning Rights in Israel.

Tali Hatuka is an architect, urban designer and author, currently based in the Department of Geography and Human Environment at Tel Aviv University. Hatuka works primarily on social, planning and architectural issues, focusing on the relationships between urban renewal, violence and life in contemporary society. She is co-editor of *Architectural Culture: Place, Representation, Body* (2005, Hebrew edition) and a recent book entitled *Violent Acts and Urban Space in Contemporary Tel Aviv* (University of Texas Press, 2010; Resling, 2008) addresses the way violent acts over the past decades have profoundly altered civil rituals, cultural identity and the meaning of place in contemporary cities.

Johan Lagae is an Associate Professor of Nineteenth and Twentieth Century Architectural History at Ghent University. He holds a PhD on colonial architecture in the former Belgian Congo (2002) and currently conducts research on urban history in Central Africa. He has authored articles in, among others, *Journal of Architecture, Third Text, Fabrications* and contributed to several Congo-related exhibitions.

Damiana Gabriela Otoiu is an Assistant Professor of Political Science at Bucharest University, Political Science Department, and researcher at Université Libre de Bruxelles, CEVIPOL (Centre for the Study of Political Life). Her research interests include the changing property relations in communist and post-communist regimes and the relationship between architecture and power during communist regimes. Her latest publications include the following: "The Jewish Property in Communist Romania (1945–1989) – Between the Soviet Model and the Spectacle of the Autonomy", in *New Europe College Yearbook, 2006–2007* (NEC, Institute of Advanced Study, Bucharest, 2009, pp. 191–238); and "Property Restitution", in *Encyclopedia of Transitional Justice*, edited by Lavinia Stan and Nadya Nedelsky (Cambridge: Cambridge University Press, forthcoming).

Guy Podoler is a Lecturer in the Department of Asian Studies at the University of Haifa. His research on modern Korea includes exploring the connections between history, memory and commemoration – especially as manifested in museums, parks and monuments – and the relationship between sport and nationalism. He is the editor of *War and Militarism in Modern Japan: Issues of History and Identity* (Global Oriental, 2009), and he has published, among others, in *Acta Koreana, Japan Focus, The Review of Korean Studies* and *The International Journal of the History of Sport*.

Marianne Rodenstein is a retired Professor of Sociology at the University of Frankfurt am Main, Germany, specializing in urban and regional sociology. She did historic and feminist research in this field and in urban planning. She edited books on high-rise buildings in German cities and on gendered spaces in different cultures. Now she does research on the intrinsic logic of cities in Germany (2008) and South Africa (since 2010).

Elena Trubina is a Professor of Philosophy at Urals State University (Ekaterinburg, Russia). Her teaching and research areas include philosophical anthropology, social theory and urban theory. She publishes on post-Soviet cultures and subjectivities. Her most recent book is *Urban Theory* (Ekaterinburg, 2008). Together with Arja Rosenholm, Irina Savkina and others, she has been part of the Finnish–Russian research project on changing values and the ways they are constructed by the Russian media. Together with Serguei Oushakine, she is co-editor of *Trauma: Punkty (Trauma: Turning Points)* (Moscow, Nezavisimoe Literaturnoe Obozrenie, 2009, in print).

Eda Ünlü-Yücesoy is an Assistant Professor, Graduate Programme in Architectural Design at Istanbul Bilgi University, Istanbul, Turkey. She is an urban planner and social geographer with research interests in public space, spatial relations and social and spatial transformation. She is the author of

Studies in Urban History (edited with Sevgi Aktüre) and *Everyday Urban Public Space: Turkish Immigrant Women's Perspective.*

Haim Yacobi is an architect and planner, and a Lecturer at the Department of Politics and Government at Ben Gurion University. His academic work focuses on the urban as a political, social and cultural entity. The main issues that stand in the centre of his research interest in relation to the urban space are social justice, the politics of identity, migration, globalization and planning. In 1999 he formulated the idea of establishing "Bimkom – Planners for Planning Rights" (NGO) and was its co-founder. His book *The Jewish-Arab City: Spatio-Politics in a Mixed Community* was published by Routledge in 2009.

Introduction

Haim Yacobi and Tovi Fenster

The chapters in this book critically explore and conceptualize how urban spaces are designed, planned and experienced in relation to the politics of collective and personal memory. The book analyses and theoretically discusses different cases in which contested national, ethnic and cultural sentiments that shape memories and practices of belonging, clash in shaping urban spaces. Examples of such situations exist in many parts of the world in which communities construct their 'past memories' (and thus their collective forgetting) within their current daily life and future aspirations. However, the diverse case studies in this book offer a discussion that goes beyond such a claim to explore how the very acts of planning and urban design (which are often considered 'rational', 'professional' and 'neutral') are rooted in the existing power structures of colonialism, modernity and nationalism in terms of knowledge production and agency. In this context David Harvey's view of modernistic planning and development has great validity. He uses the term 'creative destruction' to describe physical demolition validated by means of capitalistic reason and the modernization enterprise (Harvey 1989: 32). Sandercock (1998) builds on this definition by stating that modernistic planners are no other than 'thieves of memory', who perceive 'development' as an act of displacing communities and demolishing their houses, all in the name of progress.

Based on critical theories of planning and urban design, this book focuses on both the symbolic and tangible construction of place in cities. Looking at different case studies within North America, South Asia, Africa, Eastern Europe and the Middle East, we wish to open up an interdisciplinary debate that includes the fields of architecture, geography, planning, anthropology, sociology, urban studies and cultural studies – all areas represented by the background of our contributors. The richness of the case studies and the theoretical debates reveal some insights concerning the conflicts over memory and belonging which are spatially expressed and mediated through the planning apparatus.

The discussion of architecture and town planning as socially constructed manifestations of the state is central to this chapter. Like other cultural representations, the planned landscape is also a symbol of the political power of the state, which struggles to establish a particular collective identity and no other (Swartz 1997; Vale 1992; Yacobi 2009). As Foucault noted (1980), the rise of nationalism in Europe in the eighteenth century brought with it a significant change in the role of architecture as a tool of a new political order – one that established the state as an organization that enforces territorial, social, political and cognitive order, moulding norms and rules through mechanisms of domination, exclusion and inclusion.

Two historical examples will illustrate the above claim. The first occurred at the end of the eighteenth century in Germany with the rise of romanticism, which was an essentialist movement that considered nationality to be an emotion residing in the human psyche (*volksgeist*). This idea was supported by contemporary architectural discourse, which viewed architecture as a plastic manifestation of the human spirit and concluded that a particular architectural style – in this case, the Gothic style – could be an authentic representation of the nation. Another example is the Gothic Revival in the first half of the nineteenth century, in which the Gothic style came to represent nationality in Great Britain. This issue was often debated within elite circles, and reached its peak when the Palace of Westminster was constructed in the Neo-Gothic style as a symbol of British nationalism (Forty 1996). Marrying the Gothic style and British nationalism relied on rationalization and pseudo 'scientific-historical' facts that were described by Collins (1967) as 'an obsession for archaeology' that called upon history as a scientific discipline to prove the link to the past.

By emphasizing both 'remembering' and 'forgetting' in the production of urban space, this book follows the growing interest in understanding the substance and multiple expressions of memory and commemoration not only in their sociological or political implications but also in their spatial expressions. The very wide scope of such a discussion demands a clear focus; therefore, we have outlined the following questions to be addressed throughout the chapters: (1) what theoretical approaches concerning the interrelationships between remembering, forgetting and the production of urban space are available?; (2) what are the local political, social and cultural circumstances that frame space production in (re)constructing memory in the urban realm?; and (3) to what extent do bottom-up initiatives and the resistance of subaltern groups have the potential to transform urban space vis-à-vis top-down official planning?

Despite the historical, geographical and social differences of the cases presented in this book, some key themes can be identified that highlight the relevance of discussing memory in relation to the built landscape. The themes of nationalism, (post)colonialism, ethnicity and citizenship frame the case studies selected for this book, yet our authors examine them through the lenses of the social and spatial construction of remembering and forgetting. One could argue that the chapters in this book represent situations that characterize many cities and spaces nowadays. Our argument goes further by emphasizing the role of space design and planning in the act of remembrance and/or forgetting.

One of the central themes raised throughout the book has to do with the fact that different communities living in urban spaces struggle to legitimize their identity and diasporic memories through spatial practices. We see this, for example, in Chapter 1 by Efrat Eizenberg in which she examines community gardens in New York City as spaces for the reconstruction of a personal and cultural landscape. Through her fieldwork, Eizenberg finds that environmental memories and cultural traditions of gardeners and even the history of the space itself are being reproduced symbolically and practically in the gardens. Memories of green landscape, of other places, of farming and gardening, and of the pre-modern landscape of the area charge the experience and the representations of the gardens. Sociologically speaking, she illuminates how this reconstruction, in turn, renders urban residents into social actors – the active producers of space rather than merely its passive recipients.

In a similar vein, though in a different ethno-national context, in Chapter 2 Haim Yacobi critically examines the Western-Modern orientation of Israeli space production vis-à-vis its diverse population, whose diasporic memories often represent a material culture that does not comply with that of the nation. Yacobi's chapter focuses on the case of Netivot, a peripheral development town that offers an alternative experience of a sense of place, one linked to a diasporic-Mizrahi identity and collective memory that undermines the Israeli sovereign production of space. Theoretically, this chapter acknowledges the centrality of practices that are being conducted in the third space as a tangible site in which diasporic place is produced within a national-sovereign space.

While Yacobi's chapter refers to the experience of migrants who are part of the national project, in Chapter 3 Eda Ünlü-Yücesoy turns to the Dutch context to examine the relationship between everyday neighbourhood spatial practices of Turkish immigrant women and the social construction of urban public spaces. Ünlü-Yücesoy's ethnography in a town in the east of the Netherlands reveals the diversity of Turkish immigrant women's social constructions of neighbourhood urban public spaces, which in turn frame their sense of belonging. Through her analysis, she refers to the different characteristics and kinds of neighbourhood places, as well as their design and management, which condition the use and users' spatial interactions.

The ability to produce tangible spaces that manifest a community's collective memories is not an obvious right, as Safa Abu-Rabia discusses in Chapter 4 on the Bedouins in the Negev region of southern Israel. Abu-Rabia's chapter describes the struggle of the Bedouin to revive their past memory as an inseparable part of their exiled identity, using local strategies to preserve their historical link to their original lands. Based on ethnographic research, she examines the effect of their forced removal from their original space and their forced settlement, pointing to the ways in which the notion of exiled identity is reproduced within the second, third and fourth generations, shaping their struggle for their lands and their current sense of identity and future.

But what is the role of planning per se in giving spatial expression to the struggle over memory? This issue stands at the core of Chapter 5 by Tovi Fenster, which focuses on contradictory expressions of spatialized memories and practices of belonging among Jews and Palestinians living in the Galilee in northern Israel. This chapter examines the conflicts over planning procedures, which engage the contradictory memories that exist at both the national and local levels of planning. It explores how the dynamics of power relations can operate differently at each level and how they can result in planning resolutions that link differently to the constructions of memory and belonging of Jews and Palestinians. In order to provide a detailed analysis of the expressions of memory, belonging and commemoration, Fenster's chapter re-visits a 2002–3 planning project that she discussed elsewhere (Fenster 2007), one in which the Jewish communal village of Yaad submitted a plan for a new neighbourhood on the ruins of the Palestinian village of Miar. The chapter re-examines the developments of this planning event using the theoretical concept of therapeutic planning to show how the two communities – the Jewish and the Palestinian – deal with the situation of contrasting memories and sense of belonging and how they make efforts to bridge the gaps between these complicated sentiments.

In Chapter 6, Elena Trubina uses the term 'rhetorical space' to reflect an important aspect of the spatio-social construction of memory. This chapter discusses the nearly complete devastation that followed the Battle of Stalingrad, presenting postwar Soviet planners with a tabula rasa upon which they were expected to build an urban icon reflecting the invincibility of the Soviet regime. However, the sense of the city one gets when joining a guided tour through present-day Volgograd is rather ambivalent, leading Trubina to argue that such a fleeting experience – through which memory comes into existence at a given time and place for a particular audience – can be a useful way to deal with a location that seems to encourage one kind of utterance and performance and discourage another. This chapter allows us to emphasize the variety of city sites and visual artefacts that together produce a set of imperatives defining what can be said about the city and how it should be seen. By examining how the guided tours through the city have been conducted, Trubina proposes that Volgograd's rhetorical space works to impose on visitors a passive acceptance of a version of history that seeks to promote the smallness of the individual in the face of the larger-than-life preoccupations of the state.

Clearly, the nexus between memory and identity as mediated through the built landscape is a central theme in this book, especially in relation to nationalism. This discussion is prominent in Chapter 7, written by Guy Podoler, about the city of Seoul. In this chapter, he explores how the turbulent history of Korea and its people is reflected in the changes and developments that Seoul, which was designated the capital city in 1394, has undergone. In 1948, the city became the capital of South Korea, and the act of defining and redefining the country's national identity through intentional changes in the urban landscape has been woven throughout its process of postcolonial growth. By relying on the dichotomy between 'myth' and 'memory', Podoler highlights some of these changes. Arguing the importance of understanding Seoul as nested in multilevel configurations, Podoler explores some of the city's prominent mnemonic sites in order to shed light on the process of identity formation under changing socio-political conditions in South Korea.

In a similar direction, though in a different political and cultural context, Damiana Gabriela Otoiu analyses, in Chapter 8, the reconfiguration of urban spaces in (post-)communist Romania, focusing particularly on the issue of identity and memory in the case of the Jewish community. Otoiu's emphasis is on the legal context in which, like most Central and Eastern European countries, Romania instituted restitution policies after the collapse of the communist regime. However, as she notes, the legislation concerning the restoration of private property at the beginning of the 1990s privileged certain victims in the Central and Eastern European countries (those of the majority population), while excluding the compensation of minorities, the non-citizens (emigrants who had lost or renounced their citizenship) and the non-residents (citizens of a state who reside abroad). Within this legal context, Otoiu's research concerns the restitution of Jewish properties in post-communist Romania, whose policies are based on the concept of privileging and exclusion, as well as on the idea of restitution as an attempt to reconstruct national identity. According to Otoiu, the legislative redefinition of private property through the restitution laws and the public debates around these laws tells something about how the legislators have constructed the 'other' through the restitution policies.

In Chapter 9 Marianne Rodenstein focuses on the rebuilding of the old city of Frankfurt am Main, which was destroyed during the Second World War. In the autumn of 2007, Frankfurt's city council voted to reconstitute the core of the old city in accordance to the old street grid of the Late Middle Ages and to rebuild seven of the devastated residences and businesses. Rodenstein uses her chapter to analyse what led to the council's decision after some 63 years, during which those structures of the old city seem to have been forgotten. She also explores the reasons that they might suddenly have remembered, after so long, and the connections between city planning and remembering and forgetting. One of Rodenstein's answers is that forgetting and remembering are normal functions of memory, whereby without forgetting, which does not occur intentionally, as a rule, there can be no remembering.

Chapter 10 takes the readers to the African continent and presents a critical analysis of the built heritage in the city of Lubumbashi, Democratic Republic of the Congo. In this chapter, Johan Lagae questions the binary structure of the notion 'shared heritage'/'patrimoine partagé' that has emerged in recent debates on built heritage in former colonial territories. In the discourses of, for instance, ICOMOS, the notion stands for a heritage 'shared' by former 'colonizers' and former 'colonized', both categories being considered – albeit often implicitly – as homogenous entities. Influenced by the work of the Mémoires de Lubumbashi group, as well as recent scholarship in the field of architectural history informed by postcolonial studies, the approach on built heritage presented in Lagae's chapter is twofold. On the one hand, a plea is made to link the city's urban form to colonial history by relating it to the cosmopolitan society that produced and experienced it. On the other hand, an approach is suggested that acknowledges how specific urban places and buildings in the city are currently being re-appropriated as 'lieux de mémoire' by a variety of agents that do not necessarily (want to) share this heritage.

The last chapter, an epilogue written by Tali Hatuka, reflects upon the chapters presented in this book. Hatuka argues that collective memory becomes a tool in modifying a sense of place vis-à-vis the growing significance of collective memory.

Her reading of the chapters brings to the fore a discussion on the ways in which citizens have the opportunity to challenge the representation of future places and the way their memories will be conceived by professionals.

Finally, we hope that this book presents new insights that help planners, architects, sociologists, anthropologists and researchers in cultural studies to deeply explore and better understand the intricate and complicated relations of notions of remembrance and forgetting that are becoming so important in the development of cities today and tomorrow.

References

Collins, P. (1967) *Changing Ideals in Modern Architecture*, Montreal: McGill University Press.

Fenster, T. (2004) 'Belonging, Memory and the Politics of Planning in Israel', *Social and Cultural Geography*, 5(3), pp. 403–417.

Fenster, T. (2007) 'Memory, Belonging and Planning', *Theory and Criticism*, 30, pp. 189–212 (Hebrew).

Forty, A. (1996) "'Europe is No More Than a Nation Made Up of Several Others..." Thoughts on Architecture and Nationality, Promoted by the Taylor Institute and the Martyrs' Memorial in Oxford', *AA Files*, 32, pp. 26–37.

Foucault, M. (1980) *Power/Knowledge: Selected Interviews and Other Writings, 1972–1977*, ed. C. Gordon, Brighton: Harvester.

Harvey, D. (1989) *The Urban Experience*, Baltimore: The Johns Hopkins University Press.

Sandercock, L. (1998) *Towards Cosmopolis: Planning for Multicultural Cities*, London: Wiley.

Swartz, D. (1997) *Culture and Power: The Sociology of Pierre Bourdieu*, Chicago and London: University of Chicago Press.

Vale, L.J. (1992) *Architecture, Power, and National Identity*, New Haven and London: Yale University Press.

Yacobi, H. (2009) *The Jewish-Arab City: Spatio-Politics in a Mixed Community*, London: Routledge.

Chapter 1

Remembering Forgotten Landscapes: Community Gardens in New York City and the Reconstruction of Cultural Diversity

Efrat Eizenberg

Figure 1.1   Mott Haven Garden, South Bronx

This chapter focuses on processes of urban transformation stemming from the dialectical forces operating on space. On the one hand is the hegemonic force that strives to fully appropriate space into the system of capital accumulation by means of commodification.[1] Urban reconstruction, for example, has been generalized from the 1980s as a global strategy of urban expansion. It is a means for embedding

---

1   David Harvey, *Space of Global Capitalism: Towards a Theory of Uneven Geographical Development* (New York: Verso, 2006).

the logistics, threads and assumptions of capitalism more deeply into the urban landscape, and a powerful tendency towards the complete urbanization of the world.[2]

On the other hand is the "reaction to the vagaries of urban life"[3] and the efforts made by marginalized residents to protect their diminishing right to the city and preserve public and community spaces from being completely subjugated to the logic of the market. Space, therefore, envelops dialectical trends of power and resistance, hegemonic space and socially-produced contested space, homogeneous space and differential space that is produced or maintained as *Other* and does not conform with the hegemonic space.[4]

The "opening" for social change *through space* that is embedded within the process of *the production of differential space* can spring out of these constant negotiations over space.[5] This chapter portrays the experiences of community gardeners[6] in New York City and analyses the political development of residents that is afforded by the specific type of interaction between urban residents and their environment. Three aspects of the interaction between gardeners and their garden are discussed. First, the garden is presented as a place where individuals better recognize themselves in the physical environment through reconstructing elements of their past landscape. Memories of past landscapes that are part of the environmental autobiography of gardeners are re-enacted by gardeners in the space of the garden and in turn restore the connection of gardeners with their living environment. Second, increasingly excluded experiences of aesthetics and celebration are afforded by the gardens and offer a unique contribution to the urban experience of participants. Third, the gardens are examined as spaces for revitalization of excluded cultures that are not usually expressed in the urban space. Altogether, this chapter shows the mechanisms by which space becomes central to people's understanding of themselves and their everyday life and, as Lefebvre[7] suggests, are reintroduced to their histories and cultures. It shows how space is constituted as an arena for new practices and consciousness, the social and political significance of which goes beyond the realm of personal experience.

---

2 Henri Lefebvre, *The Urban Revolution* (Minneapolis: University of Minnesota Press, 2003).

3 Amin Ash and Nigel Thrift, *Cities: Reimagining the Urban* (Cambridge: Polity Press, 2002), 4.

4 Lefebvre, *The Urban Revolution*; David Harvey, *Spaces of Hope* (Berkeley: University of California Press, 2000).

5 Henri Lefebvre, *The Production of Space* (Oxford: Blackwell Publishers, 1991).

6 Interviews were conducted with community gardeners in New York City and with representatives of support organizations and municipal agencies. The use of quotations from interviews throughout the chapter follows the original text except for minor grammatical corrections made by the author in sentences that would otherwise be difficult to understand. The names of the gardeners that appear in the chapter were changed.

7 Lefebvre, *The Urban Revolution*.

Community gardens in New York City are green open spaces that are commonly maintained by local residents. Residents can become garden members for an annual membership fee ($10–$30 per household) and participation in workdays and membership meetings. Members receive a key to the garden and the toolshed, and in some gardens, a plot for individual cultivation (see Figure 1.1). Although varied in size and form of usage the gardens are usually used for horticulture, small-scale food production, cultural and social gatherings, and art events. There are about 650 gardens in New York City, covering together at least 90 acres of urban land and are concentrated in predominantly low income, ethnic minority neighbourhoods in Manhattan, Brooklyn and the Bronx. Most of the gardens were initiated and produced by residents in the 1970s and 1980s on vacant lots and collapsed buildings in a time of urban disinvestment.[8] They represent different ideas of urban land use in which the "spaces have rarely been planned as part of development but happen after the fact, often on deserted, derelict or otherwise unused land".[9]

In 2002, after several years of forceful public struggle against the municipality's intention to evacuated most of the gardens for private development, 400 community gardens achieved a preservation status under the City's Parks and Recreation Department and about 120 more were purchased by nonprofit organizations that sustain them as land trusts.[10] The remaining gardens are either under various ownerships[11] or are designated for future development by the Department of Housing Preservation and Development. The gardens are autonomously managed and maintained by garden members and serve as semi-public spaces; they are open to the public on weekends and whenever a garden member is in the garden. Members decide on the purpose, design and usage of the space according to their worldview, culture and needs.

Reconstruction of Past Landscapes

Memories of past landscapes and past practices are constitutive of the production of space of community gardens. Memories of natural landscapes, forests and greenery, and memories of gardens and gardening are commonly invoked by

---

8   For a detailed history of community gardens in New York City see Karen Schmelzkopf, "Incommensurability, Land Use, and the Right to Space", *Urban Geography*, 23 (2002): 323–343; Laura Lawson, *City Bountiful: A Century of Community Gardening in America* (Berkeley: University of California Press, 2005).

9   Lawson, *City Bountiful*, 2.

10   The Trust for Public Land (a national organization) and the New York Restoration Project (a local organization sponsored by Bette Midler) purchased the lots of 69 and 57 gardens, respectively, from the municipality.

11   Small numbers of gardens are currently the property of the Department of Transportation, Police, Human Resource Administration, New York State, Private Ownership and more.

gardeners. These memories propel residents to join a community garden and devote time and efforts to the creation of this communal space.

The image and experience of the garden symbolize for many gardeners their pre-urban life and in some cases their pre-United States life. As the story of Ilya, a gardener in the Bronx suggests:

> I'm from Iran ... so I think that was a big part of my connection to nature. There, in Shiraz, where I grew up that is the city of gardens and poets, the whole city is like trees and plants and parks and gardens in the middle of the house as well as outside and lots of fruit trees, those were really highly praised. We always go out and pick apples or get some walnuts and get all our hands black, or we go out and take some berries and mulberries ... so it was always a big part of our growing up, was to be part of climbing and eating and playing and natural stuff.

The symbolic reconstruction of past landscapes in the space of the garden is also accompanied by the re-enactment of past practices and experiences of space. Most of the stories of gardeners, like that of the gardener from Iran, include a variety of practices that used to be part of their everyday life in the past and are afforded again by the garden. Sam, a gardener from Harlem, Manhattan echoes this idea:

> I come from New Jersey which is the Garden State and I worked on farms and stuff like that in the past and when I was young, when I was growing up, my parents always had a garden in the back yard and we had a grape vine and peach tree and we grew collard and pepper and tomatoes and stuff like that and my mother did a lot of canning of vegetable and fruits and stuff. Because they came from the South, you know, so that was really important for them, having a stable community, taking the crops and store them. So we become self sufficient. So they pass some of that along to me and I became involved with the garden. So we planted the cherry tree, palm trees, mulberry trees, the flowers, and vegetable.

Not only does the reconstructed landscape help gardeners familiarize the space of the city to them it also allows for meaningful engagement with the space by enabling them to practice a part of their past repertoire of behaviours. Picking fruits, growing fresh produce, gardening, canning and being self-sufficient are practices that the city, generally speaking, does not support in either private or public forms. In the gardens, however, residents establish a space where these practices are not only possible but also celebrated.

The rural ethos that dominated the American geographical imagination for generations is an important factor in the powerful experience of a green oasis in the midst of a massive grey urbanity. This ethos is embedded in either the life experience of gardeners or the narratives communicated to them by parents and families. The perceived superiority of the rural environment over the urban shapes the needs and attitudes of urban residents. The words of Claudia, a gardener from East Harlem (and originally from New Mexico): "I missed gardening because I grew up gardening and New York City was hard to me because it is so urban and I grew up in a more rural environment", suggest that the gardens can address those needs by providing the opportunity to (symbolically) reconstruct past landscapes and re-enact past practices.

For the gardeners, the garden is a place where a sense of familiarity and belonging can flourish alongside the process of assimilation to life in the city. Juxtaposed to an urban environment dominated by Western principles of design, the garden makes the surrounding space more familiar. In this space, and because of it, gardeners can relate to their neighbourhoods and better recognize themselves in them. Moreover, the garden persists as a "balancing factor" that moderates the excessiveness of the urban experience by symbolically representing a remembered, more familiar and longed for open, wild, natural landscape. As Ilya from Iran suggests "and again, oxygen, I wouldn't live in New York City if there wasn't a place I could put my hands in dirt, and play, have fun, and make a living out of that and be able to laugh and enjoy myself".

Sometimes, the gardens represent not a reconstruction of past landscape but actually their negation. But even then, the gardens are clearly integrated into the environmental autobiographies of gardeners, and constituted as a completion of and compensation for something that was missing in their lives. The story of Mr Thomas, a gardener from East New York, Brooklyn, who grew up in one of the housing projects of Upper Manhattan exemplifies that point:

> And slowly I began to learn a little more because I'm not a gardener at heart.
> I don't know anything about this. I'm concrete. I'm the city guy. But I do love
> fresh fruits and vegetables and I love the earth I love the greenery I love the quiet
> ... like what we are doing now [sitting at the garden] I can do it forever. And I
> began to learn from other gardeners.

The features of the space of the garden also invoke more liminal memories, ones which were not personally experienced. The garden connects the present human-built environment with a pre-modern one – wild, pristine – by making explicit that the place has a history of its own, one which is dramatically different from the current and which is independent of human action. Mike shrewdly perceives this connection in his garden in the Lower East Side of Manhattan: "it is [the garden] an indicator of what it used to be here. The willows show that it is a wet area; it is a constant reminder that there are creeks and tidal estuary, huge tidal estuary. I would like to make that more obvious by maybe day-lighting the creek".

Martinez[12] emphasizes the gardens as a site of social history. The space, alongside photos and stories of gardeners serve as a mnemonic for the bad conditions of the site

---

12   Miranda Martinez, "The Struggle for the Gardens: Puerto Ricans, Redevelopment, and the Negotiation of Difference in a Changing Community" (unpublished dissertation submitted to the department of sociology, New York University, New York, 2002).

and the neighbourhood that used to be before the gardens were established. In addition, many gardens are used as memorial sites to commemorate loved ones or important figures from the neighbourhood or from gardeners' history and culture. These gardens are usually named after the remembered person and present commemorative objects such as sculptures, murals, photos and signs. Using the space in this way reinforces what Dagger calls civic memory: "the recollection of the events, characters and developments that make up the history of one's city or town". Civic memory is a crucial component that draws residents, or in the case of community gardens, members, together and "generate[s] a sense of civic identity".[13]

Experiencing past landscapes as an integrated component of the everyday experience of urbanity invokes and complicates the awareness of residents in regards to their space and environment. The significant component here is spatial diversity. The everydayness of the interaction allows gardeners to retain two (or more) different (sometimes contrasting) mental images and sets of attitudes and affects towards their environments. As a result, both the dominant and the alternative environments (with their complexities) become more salient. The idea that an alternative experience of the environment is possible undermines the hegemonic image of the city. The fact that the garden is their own product facilitates an understanding that they themselves can make the alternative happens. This important realization makes space and environment-related issues even more prominent in their consciousness.

Aesthetic Experience and Life as Celebration

Aesthetic and celebration are central to people's experiences of the gardens, but they are usually given a minor attention in our understanding of the experience of space and the urban experience in particular. The reason for the insufficient treatment that these two subjects are given in social sciences can be attributed to their decreased importance in modern life governed by instrumental rationality. In fact, Henri Lefebvre[14] in his impressive critique of everyday life suggests that the two are essential to the process of de-alienation of the everyday.

There is a lot to be said on the aesthetic value that gardens introduce to neighbourhoods. A variety of traits are considered to render cities "aesthetically pleasing":

> ethnic and cultural variety ... a diversity of vegetation ... public art and freedom
> of expression in the community in the forms of sculpture, graffiti and street

---

13   Richard Dagger, "Metropolis, Memory and Citizenship", in *Democracy, Citizenship and the Global City: Governance and Change in the Global Era*, ed. Engin Isin (London: Routledge, 2000), 37.

14   Henri Lefebvre, *The Critique of Everyday Life: Foundations for a Sociology of the Everyday* (New York: Verso, 2002).

art, a range of build-out (or zoning) that creates both densely and sparsely populated areas, scenic neighboring geography (oceans or mountains), public spaces and events such as parks and parades, musical variety through local radio and street musicians, and enforcement of laws that abate noise, crime, and pollution.[15]

Community gardens in New York City allow for many of these traits to materialize, and therefore contribute to the aesthetic experience of urbanites in general and gardeners in particular.

By aesthetic I refer to a sensory experience of the environment which is that of beauty and satisfaction, as well as involvement in creative activities that are perceived to yield aesthetically pleasing results. The gardens provide a unique aesthetic experience which differs from the surrounding built environment and from urban parks. They all differ in vegetation (trees, bushes, flowers, fruits, vegetables, water plants, herbs, etc.), animals (mostly birds, butterflies and bees, turtles, fish, squirrels, hens, etc.), structures (gazebos, benches, toolsheds/casitas, amphitheatres, stages, etc.), design elements (individual/common plots, elevated plots, lawns, ponds, etc.), art displays (sculptures, murals, mandalas, etc.), and art events (theatre and music performances, movie screenings, art exhibitions, etc.). Visitors to the garden can therefore engage with a wide variety of assemblages of colours, light, levels of seclusion from others, smells, sounds (birds singing, leafs in the wind, music, etc.), and sensations (cool, breezy, shady, etc.) according to their preference.

Gardeners are involved in the creation of their own aesthetic vision through the design of the landscape, assembling flowers, plants, stones and structures. It is very common to find additional decorations like stones and sculptures across the garden and as personalizers of individual plots (see Figure 1.2). The gardens exist as open creative outlets in the city for both observers and producers of art. It is a place where people can exhibit or perform their art regardless of their proficiency. Finally, the gardens are one of the only places in the city where public art – a movement that flourished in the 1980s and had since been eradicated from the city's streets[16] – can still be encountered. An interesting example is the many murals that cover the walls of the buildings in the perimeters of the gardens that express various social and political commentaries. Some deal with the struggle for land that was waged by gardeners, others deal with abstract issues such as: freedom, community, nature, etc. (see Figure 1.3).

15 Wikipedia "Aesthetics" http://en.wikipedia.org/wiki/Aesthetics#Urban_life (accessed on 19 July 2007). See, for example, Daniel Botkin and Charles Beveridge, "Cities as Environments", *Urban Ecosystems*, 1 (1997): 3–19; Hanna Mattila, "Aesthetic Justice and Urban Planning: Who Ought to Have The Right to Design Cities?", *GeoJournal*, 58 (2002): 131–138.

16 Iria Candela, *Sombras de Ciudad: Arte y Transformacion Urban en Nueva York, 1970–1990* (Madrid: Alianza Editorial, 2007).

Figure 1.2   Personalizers of individual plots

Figure 1.3   Murals[17]

This richness of aesthetic experience, being less prevalent in urban life but abundant in gardens, is not taken for granted by the gardeners. The words of Ilya, a gardener from the Bronx, poetically capture the unique nature of the gardens:

> A garden is probably the perfect art project because it works with the community, it lives by itself. I always loved art projects that when made started creating their own life. Other people work with them, somebody do this and that and it started do its own thing. But garden does that naturally it grows, people plant thing, we have sculptures, we got performances, kids grow up in it. And so it an ideal and it's all the people who are the community sharing their vision of what they believe and what they want.

The idea of celebration interacts with that of aesthetic experience. It involves being and doing for the sake of being and doing, or for the sake of joy outside of the realm of

---

17   [Top left] Lower East Side, Manhattan "The Struggle Continues". [Top right] Lower East Side, Manhattan (translation from Spanish by Breixo Viejo) "Irresistible forces/ The idea is steamed: the divine steam,/Which, invisible and powerful as the wind,/Securely marches to its immortal destiny./Who dares to stop its movement?/If a mountain rises in its way,/It opens up a tunnel to let the thought pass through/Jose de Diego". [Bottom left] Brownsville, Brooklyn. [Bottom right] East Harlem, Manhattan.

instrumental rationality. Celebration as an activity in the garden is usually manifested in parties, festivals of various kinds, street parades, neighbourhood cookouts and feed-ins, music and dancing, and various leisure workshops and programmes for kids and adults. But celebration is also manifested in the simple, less orchestrated activities such as gardening, socializing, spending playful time with one's children, finding a place of serenity, reading, and observing nature and the joyful activity of others. In short, having a place where the practice of leisure is abundant and encouraged. As Claudia, a gardener from East Harlem observed: "I think people are feeling connected to nature and it relaxes them to be in the garden. I think a lot of gardeners, a lot of them do, but some of them don't necessarily know much about gardening but do it because they enjoy being in the space and getting together".

Emily, a gardener from the East Village, provides a taste of this richness through her own experiences:

> We attend most of the workshops and musical events and we work you know to help on the events that they have, I coordinated the last Christmas party … I am very involved in that way and that is really great for me, I mean it is great for me to be able to go out with my daughter to see some excellent music for free in the evening right across the street from my house. It is very easy. There were different workshops we have attended, we attended a fish printing workshop where [my daughter] printed fish on clothing, we have attended the mandala making workshops when they were making the fence and tracing the hands 'cause these are all garden members' hands, and the hands of the children there. In fact, my daughter's hands they were very little at the time … We have attended just numerous workshops, tie-dyed cloth, making jewelry, and then I taught workshops in marble games, making rubber band balls, and making baskets.

Emily's story raises several important issues: it describes the variety of the experiences possible in community gardens, it also highlights the importance of all this abundant-ness being free, and finally it underlines the opportunity to (re)create oneself in space through celebration.

Beyond the openness of access that allows all people – the haves and the have-nots – to engage in celebration and aesthetic experience, cost-free activities lend themselves to an increased freedom of engagement. The regular rules of transaction between participants are more flexible and allow a wider spectrum of forms of participation for both performers and audience. As a result, activities are less formal and less calculated and participants can therefore take a more active part in a performance or a workshop than they could ever have in paid-for activities. In addition, these cost-free opportunities for celebration are anchored in the relative scarcity rather than the multiplicity of such experience. That is, the divergence of the experience of having-so-much-for-free from life experiences outside the garden. Where else do people have the chance to experience common wealth? What happens in the garden is not charity; the roles of givers and receivers constantly interchange and the affluence of resources exists in the artistic and social skills of the people and in the physical setting. The discrepancy between the common wealth of the gardens and the lack thereof in other realms of life is intriguing for gardeners and other participants. They are aware of this discrepancy; this not only recharges the value of the garden, it also leaves unsettled the places and activities outside of the common wealth. This unsettling experience of the world and awareness to existing discrepancies is one of the "concientizing" effects of the gardens. It is a step towards mitigating *prescription* – a behaviour of the less powerful that is prescribed by dominant guidelines.[18] The experience of an alternative challenges the prescribed dominant experience. Concientization, which entails mitigating prescription, is the process of the rethinking and rearticulating of worldviews through a critical gaze.[19]

Other issues that are raised by Emily's story pertain to collective creative abilities and (re)creating oneself in space. These issues are presented through the story of the mandala[20] that the garden members produced. Mandala-making interweaves the aesthetic and the social; "creating a group mandala is a unifying experience in which people can express themselves individually within a unified structure".[21] Emily can find the traces of her own history as it is engraved in space by her daughter's small palms on the mandala fence (see Figure 1.4). These objects that are imprinted with the people who created them serve as social mnemonics not only for their creators but also for others who encounter them.

The collective creative activities that often occur in community gardens are an instance of communal determination of space. This process of production may be understood as what Marx termed the *objectification* of species-being "for they duplicate themselves not only in consciousness, but actually in reality".[22] It is through objectification, that results from collective experience and action (or, for Marx, "positive labour") that people can "contemplate themselves in the world they have created" and at the same time become "object[s] for others within the structure of social relation[s] and in this way create civilization".[23] Objectification reverses species alienation, the separation and atomization of humans' consciousness and experience that Marx saw as the consequences of dehumanizing labour that characterizes capitalist societies. Objectification therefore is a means towards de-alienation and emancipation.

Ehrenreich suggests that the "emerging capitalist perspective [with its relentless] focus on the bottom line" presents the growing desire for "well-regulated human labor" and so leisure came to represent "the waste of a valuable resource ...

---

18   Paulo Freire, *The Pedagogy of the Oppressed* (New York: Penguin, 1972).

19   Paulo Freire, "A Few Notions about the Word 'Concientization'", *Hard Cheese*, 1 (1971): 23–28.

20   "Mandala" refers to the process of a collective creation of an art object.

21   "What is Mandala" available from: The Mandala Project, www.mandalaproject. org/What/Index.html (accessed on 19 October 2007).

22   Karl Marx, *Economic and Philosophic Manuscripts of 1884* (Moscow: Foreign Languages Publication, 1959), 114.

23   Kenneth Morrison, *Marx, Durkheim, Weber: Formations of Modern Social Thought* (Thousand Oaks: Sage, 2006), 404.

Figure 1.4   The mandala at 6&B community garden, Manhattan

festivities had no redeeming qualities. They were just another bad habit the lower classes would have to be weaned from ... recuperate from, the weekend's fun".[24] Ehrenreich stresses the importance of this repressed capacity for the generation of inclusiveness, social cohesion, happiness and integrity, and claims that it "is not easy to suppress ... the capacity for collective joy is encoded in us ... We can live without it, as most of us do, but only at the risk of succumbing to the solitary nightmare of depression".[25]

Celebration and aesthetics, therefore, in many ways stand outside of the hegemonic order of the everyday life. Together, according to Lefebvre, they are key to the process of de-alienation of the everyday life. Enjoyment of the world should not be "limited to the consumption of material goods" but driven from the rediscovery "of the spontaneity of natural life and its initial creative drive ... [and] art would be reabsorbed into an everyday". Thus, the spiritual powers of humans which are now alienated will make the "journey back to ordinary life and invest themselves in it by transforming it".[26]

---

24   Barbara Ehrenreich, *Dancing in the Streets: A History of Collective Joy* (New York: Metropolitan Books, 2006), 100–101.

25   Ehrenreich, *Dancing in the Streets*, 260.

26   Lefebvre, *The Critique of Everyday Life*, 36–37.

Gardens as Carriers of Culture

Although statistically New York City is a multicultural city with none of its many ethnic groups being the majority, the logic of space in the city, its design, usage and experience are clearly dominated by Western (white) culture. The hegemonic culture dominates and expresses itself in space, deploying mechanisms that marginalize expressions of other cultures. In spite of that, the space of the gardens is re-appropriated and used to celebrate these silenced cultures.

Most of the gardeners in New York City define themselves as Latinos (mostly from Puerto Rico) or as African Americans. Since gardens are very local sites that are established by members of the surrounding building blocks, ethnic segregation – or what Thabit defines as ghettoization[27] – explains why many of the gardens are clustered as single-ethnicity gardens (either Latino or African American). Community gardens that are located in neighbourhoods with more ethnically mixed population reflect this ethnic structure in their membership.[28] With the gentrification of inner-city neighbourhoods that had left no Italians in Little Italy, and displaced Latinos from Luisaida (now the East Village) and Spanish Harlem, there are also growing numbers of white members in community gardens.

One of the signature characteristics of the pool of community gardens, and a great source of pride for gardeners, is their variety. Each garden allows for a uniquely different experience of space with its own arrangement, aesthetic, usage, colours and so forth. This diversity is possible because gardens are the free expression of a specific group over a space without any guiding principles of urban planning and design. In the garden, gardeners can collectively express and experience their culture. Various aspects of the culture can be realized in this common space and allow for a rich, multi-layered experience that engages aesthetic and culinary preferences, rituals, customs and artistic expressions, as well as social interactions. While presenting an impressive cultural diversity, gardens could be roughly divided into three prototypes: the casita gardens; the farm gardens; and the eclectic culture gardens.

The "casita gardens" are predominantly Latino in population, and could be identified by the structure of the casita – literally a "small house" in Spanish – that "imitates traditional rural Puerto Rican homes, [the design of which] has

---

27   Walter Thabit, *How East New York Became a Ghetto* (New York: New York University Press, 2003). Ghettoization refers to the process in which intentional actions and policies prevent black population from living in white communities and force them into communities that were slated by the real estate for minority occupancy. Within these communities other policies are reproducing the marginality of this population through deprivation of services such as infrastructure maintenance, policing, education and so forth.

28   We can find ethnically integrated gardens in neighbourhoods such as Spanish Harlem (CD 11) with 36 per cent black and 52 per cent Hispanic; or in the Lower East Side that its population structure was transformed dramatically since the late 1980s with the flow of young artists and white gentrifiers into a previously Hispanic dominated neighbourhood.

Figure 1.5   9 St. community garden and park, East Village, Manhattan

been traced back to the indigenous Tainos ... [it is] brightly painted to evoke dwellings on the island".[29]

The casitas are used to store the food and musical equipment that are part of the celebration of Latino culture as well as to serve as a cosy seating place for the gardeners (see Figure 1.5). A study conducted on community gardens in the Bronx with predominantly Latino population suggests that Latino gardeners view the garden as more important for community development and as a space for social and cultural gathering over two other contributions of the gardens: preservation of open space and civic agriculture.[30]

In some casita gardens, gardeners realized the strength of the garden as a space for cultural transmission and actually transformed it to serve as a cultural centre. A successful example is the Rincon Criollo (literally: Creole Corner) Cultural Center in the South Bronx that revives the Puerto Rican working-class dances (the Bomba and Plena). This effort within the context of the casita garden is explained in the *NY Latino Journal*: "For Puerto Ricans, whose immigrant

---

29   Martinez, "The Struggle for the Gardens", 67.

30   Laura Saldivar-Tanaka and Marianne Krasny, "Culturing Community Development, Neighborhood Open Space, and Civic Agriculture: The Case of Latino Community Gardens in New York City", *Agriculture and Human Values*, 21 (2004): 399–412.

Figure 1.6   Farm garden, East New York Historic Garden, Brooklyn

experience has been one of displacement rather than assimilation, the creation of casitas like the one at Rincón Criollo, has enabled us to take control of our immediate environment and, in the process, to rediscover and reconnect with our cultural heritage".[31]

"Farm gardens" are predominantly African American in population and their space is organized around food production more than around social gathering (see Figure 1.6). In the practice of gardening and the level of self-sufficiency that it provides African American gardeners claim their culture.

In most cases farm gardens are community-oriented though the kind of engagement might be different from the casita gardens. East New York, Brooklyn, where about 50 per cent of the residents are African American (and about 40 per cent are Hispanic), is one of the poorest, and most crime-ridden communities in the borough of Brooklyn, and is also the home of the largest number of community gardens (about 60). Farm gardens in this neighbourhood reach out to the community by organizing food giveaways and community feed-ins several times during the gardening season. This might be in the form of street

---

31   Carlos "Tato" Torres, "Rincon Criollo: More than Just a Little House in the South Bronx", *NY Latino Journal*. Available from http://nylatinojournal.com/home/culture_education/ny_region/rincon_criollo_more_than_just_a_little_house_in_the_south_bronx.html (accessed on 12 September 2006).

party, or garden BBQ that is open to everyone. Another form of engagement is exemplified by the Euclid 500 Block Association community garden that joined forces with a food pantry to conduct a weekly distribution of donated food and fresh produce to the community's poor. This garden also placed a basketball hoop at its rear to attract adolescents from the streets into the garden, exchanging gardening hours for permission to play. Another farm garden in this neighbourhood is known for its many free workshops on food canning, knitting, papier-mâché hat-making and so forth.

Most prominent, though not exclusive to farm gardens, is the use of the space for the cultivation of vegetables and herbs that are part of the ethnic cuisine but are not available for purchase or are generally unaffordable. For the African American kitchen, farm gardens produce many leafy vegetables such as collard and kale, a variety of corn, and tomatoes. Casita gardens are known for their collection of hot and sweet peppers and various herbs. Mr Thomas, a gardener from East New York, Brooklyn, speaks about this aspect of gardens as carriers and educators of culture: "because there are multi-cultures in this neighbourhood some of the gardeners plant specifically for people's personal [use] ... so I get to hear about things like Kulolo and all these different things that I would never hear about if I would ... go to the supermarket to get my produce, I wouldn't know about all of this".

The "eclectic culture gardens" are characterized by a predominantly white membership and are located mostly in areas that went through or are undergoing gentrification. Some eclectic culture gardens were initiated by these newcomers, usually artists, but some were established by the non-white population before it was displaced from the neighbourhood, letting the newcomers "take over". Membership in these gardens is generally younger than in the other two types. Eclectic culture gardens, depending on their size, usually present a mixture between social space and gardening space but in general will have more areas of flowers and plants display than food production. This difference is probably related to the higher socio-economic status of members in these gardens, whereas in the other two types of gardens food production is more of a necessity. A study of gardens in the Lower East Side neighbourhood, Manhattan, an area that faced intensive gentrification since the late 1980s, finds that this type of gardens are better connected to the various green and neighbourhood organizations than the casita gardens. As such, the former has more resources to invest both in the space of the garden and in the quantity and type of events that it can offer to the neighbourhood.[32] As the name suggest, eclectic gardens feature a variety of cultures. Underlying many of these cultures is a sensibility that stretches between environmentalism to Paganism.[33] The annual Earth Day festival and the bi-annual Solstice event, celebrated in these gardens, are among the festivities that manifest these sensibilities. The events calendar of these gardens features yoga and Tai-chi

32   Martinez, "The Struggle for the Gardens".

33   Malve von Hassell, *The Struggle for Eden: Community Gardens in New York City* (Westport: Bergin & Garvey, 2002).

Figure 1.7   Eclectic culture gardens

classes, lectures on nature, eclectic music performances, and movie screenings. However, one would also find celebrations of many of the Christian holidays, as in the casita gardens and farm gardens (see Figure 1.7).

It is interesting to note at this point that the very first historical phase of community gardens in the United States was a government-initiated poor relief programme (started in 1894 in Detroit). Designed with the idea of cultural assimilation in mind, the programme served, according to its proponents, as a melting pot in which new immigrants would assume an industrious persona and learn the American way.[34] The programme was widely adopted by cities in the United States because of its financial success as a sort of "welfare-to-work".[35] However, its success in indoctrinating the immigrants and facilitating their assimilation was not evaluated. In contrast, the contemporary phase of community gardens, I argue, epitomizes the opposite mechanism for cultural assimilation. While the mechanism of melting pot de facto aims at flattening differences and assimilating into the hegemonic culture, the contemporary phase of community gardens, characterized by grassroots production of space, celebrates past experiences, and revives cultural practices and as such allows for assimilation without repression. Multiplicity of cultures then is symbolized in the physical environment of the city through the space of the gardens and is expressed in design, social interaction, food, music, dance, events and rituals.

Rooting Politics

The reproduction of forgotten landscapes, experiences and cultures exert emotional qualities and values from space which are immaterial but objective. Community

34　Thomas Bassett, "Vacant Lot Cultivation: Community Gardening in America, 1893–1978" (unpublished thesis submitted to the geography department, University of California, Berkeley, 1979); Lawson, *City Bountiful*.

35　Bassett, "Vacant Lot Cultivation".

gardens were described as a space that affords the reconstruction of past landscapes and the re-enactment of cultural praxis that are forgotten and displaced from the contemporary urban space. The opportunity to reinstitute memories (practices, aesthetics, etc.) of space in the immediate environment is an important means through which gardeners recognize themselves in the urban setting and maintain the continuity and integrity of their identity in the alienated urban sphere. The gardens become a significant space in the everyday life experience of participants and facilitate the process of attachment to the living environment. This process, that contributes to a sense of stability and belonging and as such to psychological well-being,[36] is particularly important in these neighbourhoods of New York City where 90 per cent of the residents rent their homes (rather than own them) and are threatened at some level by displacement due to gentrification.

Beyond the opportunities that the space of the everyday life offers to change one's lived experience of the city, it also encompasses the potential to act subversively and to facilitate broader social change. The everyday life is a crucial terrain of conflict between the dominant and subversive forces in society. According to Lefebvre, the split between private and public life is intensified in modernity and causes the separation of cultural and historical events from everyday life experience. This separation exacerbates the colonization of the everyday life by the dominant force and its subdued by the spectacle. Therefore, the collective reintegration of socio-cultural events to the daily experience confronts and reverses these effects and can lead to the de-alienation of everyday life.[37]

The realm of collective memories and cultural symbols of community gardens serves as a common resource for urban residents. The space of the garden is a means for gardeners to construct themselves as meaningful participants in society and is therefore a space for the traditionally marginalized to engage with the production of urban space according to their memories, cultures and identities, and become active producers of space rather than merely its passive users. Furthermore, in the process of producing their environment, community gardeners develop as political actors who strive to instate their vision of the city beyond the boundaries of the garden. Their acquired social role entails their engagement with the discourse and practices of struggle against the structural pressures to neoliberalize the urban space that repress spatial expressions of alternative memories and cultures. The production of space of the gardens should therefore be understood as both means and ends of the process whereby residents develop their political consciousness and act upon it.

---

36 Irwin Altman and Setha Low, *Place Attachment* (New York: Plenum Press, 1992).

37 Lefebvre (*The Critique of Everyday Life*) sees the everyday life as the terrain for a movement to grow and transform the prevailing system. For example, Lefebvre is fascinated with festivals as an act of production of space for creative cultural celebrations. However, the main difference between community gardens and festivals is the temporality of the latter.

References

Altman, Irwin, and Setha Low. *Place Attachment*. New York: Plenum Press, 1992.

Ash, Amin, and Nigel Thrift. *Cities: Reimagining the Urban*. Cambridge: Polity Press, 2002.

Bassett, Thomas. "Vacant Lot Cultivation: Community Gardening in America, 1893–1978". Unpublished thesis submitted to the geography department, University of California, Berkeley, 1979.

Botkin, Daniel, and Charles Beveridge. "Cities as Environments". *Urban Ecosystems*, 1 (1997): 3–19.

Candela, Iria. *Sombras de Ciudad: Arte y Transformacion Urban en Nueva York, 1970–1990*. Madrid: Alianza Editorial, 2007.

Dagger, Richard. "Metropolis, Memory and Citizenship". In *Democracy, Citizenship and the Global City: Governance and Change in the Global Era*, ed. Engin Isin, 25–47. London: Routledge, 2000.

Ehrenreich, Barbara. *Dancing in the Streets: A History of Collective Joy*. New York: Metropolitan Books, 2006.

Freire, Paulo. "A Few Notions about the Word 'Concientization'". *Hard Cheese*, 1 (1971): 23–28.

Freire, Paulo. *The Pedagogy of the Oppressed*. New York: Penguin, 1972.

Harvey, David. *Spaces of Hope*. Berkeley: University of California Press, 2000.

Harvey, David. *Space of Global Capitalism: Towards a Theory of Uneven Geographical Development*. New York: Verso, 2006.

Hassell, Malve von. *The Struggle for Eden: Community Gardens in New York City*. Westport: Bergin & Garvey, 2002.

Lawson, Laura. *City Bountiful: A Century of Community Gardening in America*. Berkeley: University of California Press, 2005.

Lefebvre, Henri. *The Production of Space*. Oxford: Blackwell Publishers, 1991.

Lefebvre, Henri. *The Critique of Everyday Life: Foundations for a Sociology of the Everyday*. New York: Verso, 2002.

Lefebvre, Henri. *The Urban Revolution*. Minneapolis: University of Minnesota Press, 2003.

Martinez, Miranda. "The Struggle for the Gardens: Puerto Ricans, Redevelopment, and the Negotiation of Difference in a Changing Community". Unpublished dissertation submitted to the department of sociology, New York University, New York, 2002.

Marx, Karl. *Economic and Philosophic Manuscripts of 1884*. Moscow: Foreign Languages Publication, 1959.

Mattila, Hanna. "Aesthetic Justice and Urban Planning: Who Ought to Have The Right to Design Cities?". *GeoJournal*, 58 (2002): 131–138.

Morrison, Kenneth. *Marx, Durkheim, Weber: Formations of Modern Social Thought*. Thousand Oaks: Sage, 2006.

Saldivar-Tanaka, Laura, and Marianne Krasny. "Culturing Community Development, Neighborhood Open Space, and Civic Agriculture: The Case of Latino Community Gardens in New York City". *Agriculture and Human Values*, 21 (2004): 399–412.

Schmelzkopf, Karen. "Incommensurability, Land Use, and the Right to Space". *Urban Geography*, 23 (2002): 323–343.

"Tato" Torres, Carlos. "Rincon Criollo: More than Just a Little House in the South Bronx". *NY Latino Journal*. Available from http://nylatinojournal.com/home/culture_education/ny_region/rincon_criollo_more_than_just_a_little_house_in_the_south_bronx.html (accessed on 12 September 2006).

Thabit, Walter. *How East New York Became a Ghetto*. New York: New York University Press, 2003.

Chapter 2

Memory, Recognition and the Architecture of a Diasporic Place: The Case of Netivot, Israel[1]

Haim Yacobi

Space, Place and the Third Space

To a visitor arriving in Netivot, an Israeli Development Town located in the peripheral Negev region, the city looks like many other development towns in Israel that conform with the modernist form of space. The housing blocks and the semi-detached houses are in accord with the road system that marks the planned neighbourhood units. This urban scheme, designed by Tzion HaShimshoni,[2] is based on modernistic planning principles such as zoning (i.e. the divided location of urban functions); green open, public spaces linked to pedestrian paths; and efficient road systems that link between the different zones.[3] Nevertheless, this schematic urban morphology is visually and spatially disrupted by indications of a different layer of urban life and experience that reflect diverse perceptions of what constitutes a city – an approach that highlights the way in which "people are never passive recipients of external initiatives, but rather always struggle within their own immediate contexts of constraints and opportunities to produce meaningful life with their own particular values and goals".[4]

Following this observation, this chapter critically examines the Western-Modern orientation of Israeli space production vis-à-vis its diverse population, that in many aspects represents material culture which does not comply with the

---

1   This chapter was originally published in: Haim Yacobi, "From State-Imposed Urban Planning to Israeli Diasporic Place: The Case of Netivot and the Grave of Baba Sali", in Julia Brauch, Anna Lipphardt and Alexandra Nocke (eds) *Jewish Topographies: Visions of Space, Tradition and Place* (Aldershot: Ashgate, 2008), 63–82.

2   David Zaslevsky, *A Survey of Netivot's Development* (Jerusalem: Ministry of Housing, 1969).

3   For a detailed discussion see Zvi Efrat, *The Israeli Project: Building and Architecture 1948–1973* (Tel Aviv: Tel Aviv Museum of Art, 2004), 807–824 [Hebrew].

4   Goh Beng-Lan, *Modern Dreams: An Inquiry into Power, Cultural Production and the Cityscape in Contemporary Urban Penang, Malaysia* (Ithaca, NY: Cornell, Southeast Asia Program, 2002), 202.

national supremacy. Specifically, this chapter will focus on the case of Netivot, a peripheral development town which offers an alternative experience of sense of place, linked to the diaspora and Mizrahi[5] identity that subverts the Israeli hegemonic production of space by creating a hybrid place.

The very particularity of the Israeli spatial reality calls for a localization of such theories towards the meaning of the built environment. The "Israeli place", as I will elaborate, is the product of a contested socio-historical process, characterized by the motivation for controlling national space and framing it in a total manner. Such a decisive approach generates counter-products which are also spatially expressed. The methodological roots of my claim originate from the tendency of urban research in Israel to focus on formal processes of space production, dictated from above and burned onto the collective mind by means of plans, thus reproducing the perception of what a place is and which sites do not warrant being called places. Connected to this debate is the centrality of the argument that the production of Israeli-Zionist space can be understood along three axes: the denial of the Orient, the rejection of the bourgeois, and the invalidation of diaspora culture.[6] However, vis-à-vis the short theoretical notes that open this section, let me propose that such an argument is partial since it refers to place production from above, ignoring the fact that vast parts of the built environment in Israel do not comply with standard regulations (legally as well as architecturally) and thus they penetrate into the spatial order created by the national culture and by so doing produce hybrid places.

It is important to clarify, already at this stage, that the notion of hybridity accordingly is not a third concept which relieves the tension between cultures, hence resulting in the recognition of the subordinate culture by the hegemonic one.[7] Rather, it is formulated within a third space – a discursive junction in which the sovereign and the colonial subject are not exclusive alternatives, and the construction of their identities involves "mutual contamination".[8] During this process, which involves mutual reproduction and imitation within the intervening, namely third, space, the colonial power also produces its alien. Therefore, claims Bhabha, the third space is potentially a site of resistance, undermining the polar perception which poses identities as opposite, authentic, ethnically and racially essentialist entities, hence it can be perceived as a site of struggle and negotiation.[9] The significance of Bhabha's argument is its recognition that power relations are the basis for the production of subaltern culture, and it proposes a wide sociological understanding of the range

---

5   Mizrahi Jews, Mizrahim in plural, are those who come from Arab and Muslim countries.

6   Alona Nitzan-Shiftan, "Whitened Houses", *Theory and Criticism*, 16 (2000).

7   Homi Bhabha, *The Location of Culture* (London and New York: Routledge, 1994), 113–114.

8   Homi Bhabha, *The Location of Culture*, 113–114.

9   Homi Bhabha, "The Third Space: Interview with Homi Bhabha", in Jonathan Rutherford (ed.) *Identity, Community, Culture, Difference* (London: Lawrence and Wishart, 1990), 211.

between the top-down power and the voice of subaltern subject. This insight, I would suggest, is an appropriate vehicle for examining space production in social and cultural theory in general and in the Israeli case in particular.

At the core of criticism against Bhabha's third space conceptualization, emphasizing the discursive aspect lays the material question. This critical tone appears both in relation to the distinction between politics and discourse and in the call for the examination of hybrid spaces within postcolonial contexts of specific geography, history and economics. Postcolonial studies have focused on textual and literary studies, being only vaguely concerned with what happened. In the context of these criticisms, there is a necessity to engage in material practices, actual spaces and real politics that have increasingly, if belatedly, brought into the debate recent as well as earlier studies of colonial urbanism and architecture, largely ignored by the literary discourses on the postcolonial context (King 2003: 167). Indeed, it is important to ground the formulation of the third space in meaningful practices; hybrid places are the result of interaction that are located in concrete differing positions of power, which must nevertheless cohabit.[10]

This chapter joins that call; it aims to acknowledge the centrality of practices that are being conducted in the third space as a tangible site where diaspora place is produced within a national-sovereign space. To put it differently, the discussion concerning the Israeli place, on which I will focus, allows for recognizing the significance of practices occurring within the third place, not merely as a metaphor, but also as a concrete site in which material practices, producing the physical space, are activated. My argument follows Raz-Karkotzkin's suggestion that disavowing the diaspora past in the Israeli context is part of the implementation of the regime of modernity. It should be recognized that a strict denial of the diaspora exists owing to the formulation of cultural identity in terms of disavowal, and more importantly, the fact that repudiating the diaspora means repudiating the Jewish memory.[11]

Past, Future and the Jewish Place

The general debate concerning the Israeli production of space-place focuses on the "local" versus the "other", ignoring the dynamic nature of identities. The same applies to the examination of the role of architecture and planning within the Zionist enterprise, which has so far been focused on the Jewish-Arab or Israeli-Palestinian issue. This is connected to the very common point of departure – 1948 – to the discussion of the appropriate spatial form of habitat for the Jewish people in their motherland. Yet, it is important to follow this debate from an earlier period

---

10   Nezar AlSayyad, "Hybrid Culture/Hybrid Urbanism: Pandora's Box of the 'Third Place'", in Nezar AlSayyad (ed.) *Hybrid Urbanism: On the Identity Discourse and the Built Environment* (Westport: Praeger, 2001).

11   Amnon Raz-Karkotzkin, "Exile within Sovereignty: Towards a Critic of the 'Negation of Exile' in Israel Culture", *Theory and Criticism*, 4 (1993), 113 [Hebrew].

– the pre-state period – as I will discuss in the following section by means of an analysis of the architectural discourse in the 1930s as presented in the first issues of a journal named *HaBinyan*.[12]

The journal exposes a serious discussion concerning the commitment of architects and planners to defining the appropriate form of Jewish habitat and its relevance to the construction of identity and the attempt to prove territorialization. In the first issue of *HaBinyan* there is a clear expression of the tension between the Western approach to planning and the local geographical and economic conditions in Eretz Israel:

> In the course of our adaptation to the conditions of the Land we learned …
> that neither American or European models of development, even the most
> progressive of them, are not appropriate to our capability … since in the future
> they will cause the growth of public expenditure.[13]

For Posner, it is significant to adopt a neutral modernist attitude. Thus, he argues that, "lately Jews have taken part in the development of European taste". This is expressed by the fact that the Jewish people "are distancing themselves from traditional forms, they are learning to appreciate cleanliness and simplicity, and are thus liberating their homes from the memories of the past". This liberation from the past has a considerable impact upon the presence of Jews in the Land of Israel – not just as the denial of the diaspora, but also as the denial of the Orient that is presented by means of Oriental morphology:

> First of all, people are no longer captivated by the Oriental appearance. Anyway,
> we have relinquished the Oriental character created from constructing domes
> and arcades. This reaction is necessary as well as suitable to the real demands
> of Jewish taste.[14]

The third issue of the journal that deals with villages in the Land of Israel draws attention to the dichotomous attitude towards the Oriental-Arab landscape. In the opening essay, Posner categorizes settlements and cites their disadvantages and merits. He suggests that the village in the Land of Israel "is ancient and has hardly changed". Thus, he asks "what we can learn from such ancient experiences? Probably we can learn from their economy, their social relations, their collective agricultural manners … Some people claim that the home in the Arab village protects one from the climate better than our homes in the *moshavot*".[15]

---

12   "The Building" in Hebrew.
13   Avraham Schiler, "Land Development Problems for Housing", *HaBinyan* (1937), 28–29, translated by the author.
14   Posner Yaakov, "The Village in Eretz Israel", *HaBinyan* (1938), 1, translated by the author.
15   Yaakov, "The Village in Eretz Israel", 1, translated by the author.

Beyond such an Orientalist approach, Posner continues to argue that the journal equally avoids romantic superlatives concerning the wholeness of the Arab agricultural villages, stating that "we would not say that we must build so traditionally and we would also say it is prohibited to build so badly and oddly. The Arab village is not a model for replication by us".[16] This approach indicates the centrality of architectural discourse and space production to identity. More specifically, it reveals the duality in relation to the Oriental landscape – it is on the one hand an authentic object of desire that might inspire the shape of habitat of the Jewish people, and on the other hand it is the signifier of the underdeveloped Oriental-Arab society. Yet the spatial implementation of such an approach was limited. As I will discuss in the following section, it was only when the geopolitical conditions had changed and the Israeli state was established that a modernist paradigm in planning and architecture became central to the production of the new Jewish place.

But which landscape was supposed to replace the built environment that was marked as Oriental? The answer to this question was obvious at the time and can be related to the modernization project that provided justification for the rejection of an Oriental past and present as I have discussed above. In addition, it should be related to the manner in which power relations enable the implementation of a plan aimed at providing the means to create social transformation. Realizing this plan requires the extensive involvement of the state and a centralistic planning approach to enable the fulfillment of a vision that provides "an opportunity to rewrite the national history".[17]

Modernity and urbanism in this sense are not part of an uncontrolled evolutionary process. Rather, as sociological and political processes, they crawl along and in most cases erupt via their various agents – settlement, nationalism, immigration, professional experts and capital – guaranteeing a change in society and consciousness that will eventually lead to an inevitably better future.[18] Indeed, modernity as a social project includes a doctrine of progress accompanied by the creation of a new subject, the agent of modernity, who is freed from the bonds of tradition in order to fulfil himself as an individual. Seeing modernity as a neutral process strengthens the dominance of Western culture and transforms it into the sole default option to which to aspire in order to justify being termed "modern", and this is a condition for benefiting from the distribution of rights and goods.

Let me illustrate the above argument while using one of Israel's iconic architectural objects – the *shikun* (the Hebrew word for tenement housing block), whose political and architectural meaning has been the subject of many papers,[19] indicative of its dominance in the Israeli landscape of development. Apart from being a manifestation

---

16   Yaakov, "The Village in Eretz Israel", 1, translated by the author.

17   James Holston, *The Modernist City – An Anthropological Critique of Brasilia* (Chicago and London: University of Chicago Press, 1989), 5.

18   Charles Taylor, "Two Theories of Modernity", *Public Culture*, 11 (1) (1999).

19   See for example: Hadas Shadar, "Between East and West: Immigrants, Critical Regionalism and Public Housing", *The Journal of Architecture*, 9 (2004); Rachel Kallus

of a certain school of planning, it is also linked to an ideology that perceived the formation of modern space as a means of constructing a sense of collective belonging. The *shikun* in its modernistic form assumed a double function in the Israeli context: it reflected sovereignty over national territory, and at the same time served as an incentive to economical, social and identity production and reproduction.[20]

Indeed, the construction of the *shikunim* during the 1950s was considered revolutionary. The project, which was part of a comprehensive national plan for spatial development in Israel conducted by Arieh Sharon, head of the Planning Division of the Prime Minister's Office, presumed to provide housing for a population that doubled in size during the first decade of the state.[21] The plan, entitled "Physical Planning in Israel", reflected the centralistic statehood that characterized the Israeli regime in the 1950s. The Sharon Plan defined three dimensions of spatial design (and in my opinion, a pedagogic objective as well) – land, people and time – as a basis for a professional physical construction plan. These imaginary concepts facilitated the formation of the new national space:

> This assorted immigrants' ingathering will become uniformly consolidated only if supported by comfortable physical, social and economical conditions ... A social composition and a planning framework should be provided in order to facilitate assimilation and stimulate the process of integrating different types of settlers ... into one unified creative whole.[22]

This approach enhanced the importance of the home as a vehicle for the creation of a collective sense of identity and belonging, as a means of transforming immigrants into locals; or in spatial terms, to produce place in the new territory. Golda Meir confirms this claim by stating that:

> inadequate accommodations are evident everywhere around the globe. In Sweden, no Swedish born individual whose ancestors resided there will cease to be Swedish just because he has no home. Here, however, this is severely problematic. The housing problem is highly significant, and it will determine whether that same family that emigrated with its children, foreign and unacquainted with the language, the conditions and often also the goals – it will determine whether these family members will become Israeli or remain foreign, albeit holding Israeli citizenship.[23]

---

and Hubert Law-Yone, "National Home/Personal Home: Public Housing and the Shaping of National Space in Israel", *European Planning Studies*, 10 (6) (2002).

20   Rachel Kallus and Hubert Law-Yone, "National Home/Personal Home".

21   For a detailed study see: Smadar Sharon, *To Built and be Built: Planning the National Space in the Formative Years of Israel*, thesis submitted at Tel-Aviv University (2004) [Hebrew].

22   Sharon Arie, *Physical Planning in Israel* (Jerusalem, 1951), translated by the author.

23   Golda Meir as cited in David Zaslevsky, *Housing for Immigrants – Construction, Planning and Development* (Tel Aviv, 1954), translated by the author.

The necessity for domesticating the immigrants' culture coincided with the modernist approach of planning that took it upon itself to design the housing unit. This fact had social implications that came to bear over the everyday use of private space, since it aimed at liberating the family from its traditional domestic perceptions. This planning and architectural paradigm dovetailed with the objectives of the Israeli regime in the 1950s that aimed to transform the immigrants through a process of de-Arabization.[24] Architectural modernism can therefore be contained within the parcel of national belonging under the guise of civil and secular culture – terms that according to Homi Bhabha were exploited to draw people into the human community, but at the same time were used to exclude them from it as "others".[25] These clearly reflect the double mechanism that produced the new habitat in Israel – a modernistic approach of efficiency, order and planning, and the application of ethno-national logic which replaces what has been considered as underdeveloped.

It seems that Amos Oz's description encapsulates the transformation of the *shikun* – from a pedagogical architectural object:

> The large distance between the buildings, planned by the architect, make the shabbiness more marked than it would be if the buildings were close together – a Mediterranean town, house touching house, spaces of more human proportions. Were these neglected lots intentional, in the planners imagination perhaps, meant to be vegetable gardens, small orchards, sheep pens, and chicken coops: a North African *nahalal* on the rocky slopes of Judea? What did the town planner know or want to know about the lives, the customs, the heart's desires of the immigrants who were settled here? Was he aware of, or partner to, the philosophy prevailing in the fifties that we must change these people immediately – remake them completely – at all cost?[26]

As noted by Oz, North Africa has been one of the main origins of immigration to Israel since the establishment of the state. The new national project referred to the Oriental immigrant culture as an object that demands special treatment by pedagogic Westernization and modernization aimed at reshaping the immigrant's everyday life. This was the contribution of the architectural practice and discourse to affixing antinomies such as East/West, third world/first world, modernity/backwardness and sovereignty/diaspora.

Definitely, images of the tenement housing block, as a signifier of the Mizrahim, appear in several representations that deal with the Mizrahi culture, political activism and protest.[27] Over the years, the users have transformed their housing environment.

---

24   Yehuda Shenhav, *The Arab Jews: A Postcolonial Reading of Nationalism, Religion, and Ethnicity* (Stanford: Stanford University Press, 2006).

25   Homi Bhabha, "The Third Space: Interview with Homi Bhabha", 211.

26   Amos Oz, *In the Land of Israel* (Tel Aviv: Am Oved, 1983), 28–29.

27   Haim Yacobi, "The Third Place: Architecture, Nationalism and Postcolonialism", *Theory and Critique* (2007) [Hebrew].

These additional constructions, which are not the product of professional logic and aesthetics, undermine the power of national logic, supported by professional knowledge. In other words, the modification of the housing environment is a counter-act of place determination that goes beyond the inhabitants' motivation to improve their physical quality of life, but rather as a manifestation of their past cultural affiliations – a debate to be discussed in the following section.

Towards a Diaspora Place?

> In the year 1956 the first settlers arrived at Netivot from the Maghreb countries. The Olim [new immigrants] were loaded on trucks and taken in the middle of the night to the place, the object of their yearning. Many of them believed that they were taken to Jerusalem, but under cover of darkness they were transported to the town of Netivot.[28]

The above quotation narrates in a nutshell the re-territorialization of Israel and the attempts to stabilize its sovereignty by establishment of the new development towns.[29] As part of the physical plan for Israel discussed in the previous section, the town of Netivot was established in 1956, as a regional center for the northwestern Negev agricultural settlements and the first wave of Netivot's inhabitants was primarily characterized by Jewish emigrants from North Africa. Several reports since the establishment of the city point to its economic underdevelopment, attributing this to its ethno-demographic composition.[30] Even more recent data from the Central Bureau of Statistics[31] indicates that the city is ranked at socio-economic level 3 (out of 10). In the year 2000, Netivot was officially declared a city by the Ministry of Interior and its population, as of 2004, stands at 26,000 inhabitants:[32] 70 per cent of them are Mizrahim and 25 per cent are Russian migrants.[33]

Jerusalem Street is the main entrance to Netivot – other streets branch out from it, bearing names of Jewish Moroccan saints and rabbis. At the main streets' corners, on top of the official street signs, an additional placard is placed with an

---

28   Netivot Municipality website: www.netivot.muni.il.

29   Erez Tzfadia and Oren Yiftachel, "Between Urban and National: Political Mobilization among Mizrahim in Israel's 'Development Towns'", *Cities*, 21 (1) (2004).

30   David Zaslevsky, *A Survey of Netivot's Development*; David Zaslevsky, *Housing for Immigrants – Construction, Planning and Development* (Tel Aviv, 1954) [Hebrew].

31   Central Bureau of Statistics, *Characterization and Ranking of Local Authorities According to the Population's Socio-Economic Level in 2001* (Jerusalem: Central Bureau of Statistics, 2004).

32   Central Bureau of Statistics, *Characterization and Ranking of Local Authorities According to the Population's Socio-Economic Level in 2001*.

33   According to the available data 22 per cent of the population were born in Africa while 68 per cent were born in Israel, mainly a second and third generation of the Mizrahi newcomers to the city (www.netivot.muni.il/#Top).

image of Rabbi Yisrael Abuhatzeira known as the Baba Sali (Praying Father in Moroccan Arabic) who was born in Morocco, immigrated to Israel in the 1950s and several years later settled in Netivot (see Figure 2.1).

Figure 2.1 **A placard with the logo of the Baba Sali Institutions at the entrance to Netivot**

**Figure 2.2   Pilgrims on the *Hillulah* day of commemorating the Baba
Sali's death**

Following his religious, spiritual and political influence, with the subsequent
construction of such sites, they have also gained political importance at the
municipal as well as the national level.[34] Netivot has also begun to attract Jews
who have returned to their religious roots.

The Baba Sali died in 1984. His funeral in Netivot's cemetery drew an estimated
100,000 people. The influence of the Baba Sali has grown and the city has become
a renowned focus of pilgrimage for the Moroccan Jewish community in Israel as
well as from abroad, i.e. for Moroccan Jews residing in France and Canada. His
gravesite in Netivot has become a popular pilgrimage site in Israel, especially on
the date commemorating his death (see Figure 2.2).

It is important to mention Kosansky's remark that the Jewish pilgrimage
shares similarities with saint veneration as practiced by both Muslims and Jews
in Morocco.[35] Though several anthropologists have extensively written about
this phenomenon, exploring the cultural, social and political dimension of it,[36] no

---

34   Eyal Ben-Ari and Yoram Bilu, "Saint's Sanctuaries in Israeli Development Towns",
in Ben Ari Eyal and Bilu Yoram (eds) *Grasping Land: Space and Place in Contemporary
Israeli Discourse and Experience* (Albany: State University of New York Press, 1997), 246.

35   Oren Kosansky, *All Dear Unto God: Saints, Pilgrimage, and Textual Practice in
Jewish Morocco* (Ph.D. dissertation, University of Michigan, 2003), 553.

36   Yoram Bilu and Ben-Ari Eyal, "The Making of Modern Saints: Manufactured
Charisma and the Abu-Hatseiras of Israel", *American Ethnologist*, 19 (4) (1992).

special attention has been given to the spatial influence of it, nor to its contribution to creating Netivot as a place – a void that the following section aims to fill.

If a visitor to the city were to follow the signs along Abuhatzeira Street, s/he would begin to recognize a different architectural expression of the buildings – contradicting the modernist space and commemorating the past of the Jewish community in the diaspora. This issue was raised in an interview with the representative of the Baba Sali Institutions[37] who claimed that the use of such architectural style that "purposely does not fit the Netivot cityscape ... is the appropriate way to commemorate the *Tzadik*[38] ... The buildings in Morocco in the Tafilalt region [an oasis in the Moroccan Sahara] are similar. We replicated them here in Netivot, in order to symbolize the past".

Down the road, Abuhatzeira Street leads to the edge of the city, where the modernist housing blocks mark the end of the urban constructed area. The back of these buildings faces a neglected open space, which according to the planning regulations detaches the city from its cemetery. Nonetheless, Netivot's cemetery is not a dead place – the Baba Sali burial site has become a focal point of religious, spiritual and social encounters, especially at the time of the *Hillulah*, or celebration day. From the architectural point of view, the place constitutes an attempt to establish an icon that commemorates not only the Baba Sali legend but the memory of the Jewish community in Morocco as well. This notion was raised by several people during the last *Hillulah* when I asked them what the significance of Netivot is for them. A man in his fifties told me that he was a child when his family immigrated to Israel from Morocco: "I do not remember myself what it was like there, but we come here every year with my mother ... She has told us that it is exactly the same. I feel as though I were there".[39] Let me suggest that here lies the notion of the diaspora experience for those who are here but are still attached to there.

Interestingly enough, the modification of the cemetery into a pilgrimage site has been done by official planning procedures. The new urban scheme that has been authorized enables the modification of land use from a cemetery into a pilgrimage site: "the objective of plan No. 103\03\22 is ... (c) Altering the existing land use from public, open space into a burial plot of 4,339 square meters".[40] Moreover, the modified urban scheme of the cemetery acknowledges the pilgrims' needs according to their tradition and allocates space for the construction of a "feast shelter" for use by the pilgrims, the establishment of a structure for commercial activity and the construction of three rest units:

There is a custom among some ethnic groups that the terminally ill seek healing at saints' graves by praying and seclusion, as well as by adjacent sleeping

---

37   Interview with the representative of the Baba Sali Institutions in Netivot, 25 September 2006; translated by the author.

38   Tzadik is a righteous person.

39   Informal interview, January 23, 2007; translated by the author.

40   Urban scheme No. 103\03\22; translated by the author.

accommodations. The purpose of the rest units is to enable these people to realize their wishes under the same roof [as the other activities].[41]

Spatially speaking, on the day of the *Hillulah*, the neglected space between the edge of the city and the cemetery is transformed into a meeting place of the pilgrims. Thousands of people visit the Baba Sali burial site and a lively market of religious goods, food and clothing serves the crowds (see Figure 2.3).

The extensive city life takes place in a public space, which is actually a parking lot, while the modernist *shikunim* that house many of the Baba Sali's community members serves as the backdrop. The modernist urban order is further transformed by means of the cemetery. The new religious and educational institutions that have been established by the Baba Sali Foundation are designed with reference to the "old-new" architecture and are used as landmarks on the urban scale (see Figure 2.4).

The effect of the diaspora-Mizrahi religious notion of the city is acknowledged by the municipality that participates (in terms of budget) in the *Hillulah* events and acknowledges the contribution of the institutions to the city, stating that "On its 50th anniversary, all Netivot's inhabitants appreciate the contribution of the Baba Sali to the development and progress of the city. The municipality is committed to act, by all means, in order to commemorate the Baba Sali legend

Figure 2.3   The market of religious goods, food and clothing near Netivot's cemetery

41   Urban scheme No. 103\03\22; translated by the author.

Figure 2.4   One of the new educational institutions of the Baba Sali
**   Foundation**

and to support its institutions".[42] The reconstruction of Netivot's Moroccan-like sense of place coincides with Oren Kosansky's observation of the considerable Jewish element of the city in Morocco in the mellah (the segregated Jewish quarter in Morocco) and its Jewish cemetery. Architecturally speaking, he suggests that there is a specific Jewish architecture expressed in the mellah. The main road is lined with exceptionally elevated buildings with distinct decor marking their elevation in the front. Also, he observed that the balconies extend beyond the lanes below – an element not to be found elsewhere in the old city.[43] Moreover, the interim space created in Netivot has also constructed a virtual network of places. Praying and imploring of the Baba Sali is possible via several Internet sites which also broadcast the burial site and prayers 24 hours a day.[44] I would suggest that Netivot appears to offer some insight into the potential of diaspora communities within a forceful national context to express, and often glorify, their ties to their Moroccan homeland.

---

42   Yehiel Zohar, Mayor of Netivot in the Baba Sali Foundation brochure 2006.

43   Oren Kosansky "Reading Jewish Fez: On the Cultural Identity of a Moroccan City", *The Journal of International Institutes* (8.3.2001), www.umich.edu/~iinet/journal/regional.

44   See, for example: www.po-ip.co.il.

Let me elaborate on the relevance of the notion of diaspora to our case. The concept of diaspora has traditionally referred to cases of communities living outside of their homeland. The term is used extensively while referring to emigrants, expellees, alien residents and ethnic minorities. Beyond the different definitions, there are shared characteristics of the notion of diaspora as referring to a given social group that has dispersed from its territory to a different, foreign region. In this process, the specific group constructs its collective memory concerning its origin, location, history and culture. More importantly, Safran argues, the diaspora group perceives itself as an excluded social entity which cannot be fully integrated in the host society.[45]

The definition above stems from the very specific case of the exile of the Jews from the Holy Land and their dispersion throughout several areas of the world.[46] Such an approach must be seen as an ideal type since a comparison to other cases demonstrates that most diasporas do not match this definition. In contemporary postcolonial literature, there is a wider understanding of the term that broadly refers to it in relation to displacement, dislocation, and reformation of the "double consciousness" of being "inside and outside".[47] Indeed, diaspora discourse "is loose in the world, for reasons having to do with decolonization, increased immigration, global communications, and transport – a whole range of phenomena that encourage multi-locale attachments, dwelling, and traveling within and across nations".[48]

The notion of homeland is an inherent component and used as a *raison d'être* for the production of space (as presented in the previous sections) which is related to the ideological and political circumstances that caused these people to immigrate to Israel, while at the same time, Israeli-Zionist ideology denies the notion of the diaspora past, geographies and culture of these immigrants. More specifically, as already suggested by Levy, though Jews perceive the Land of Israel as the core of their collective history, they furthermore "conceive of Morocco as a symbolic center; a homeland for those who remained behind as well as for those who migrated".[49] Likewise, as in other cases, the role of religion in the diaspora experience and place-making is central; places accumulate meaning beyond their function for religious practices and thus gain value and become social, cultural and political signifiers of diaspora identity.[50]

---

45   William Safran, "The Jewish Diaspora in a Comparative and Theoretical Perspective", *Israel Studies*, 10 (1) (2005).

46   Safran, "The Jewish Diaspora in a Comparative and Theoretical Perspective".

47   André Levy, "Center and Diaspora: Jews in Late-twentieth-century Morocco", *City & Society*, 13 (2) (2001).

48   James Clifford, *Routes: Travel & Translation in the Late Twentieth Century* (Cambridge, MA: Harvard University Press, 1997), 249.

49   Levy, "Center and Diaspora: Jews in Late-twentieth-century Morocco", 245.

50   John Fenton, *Transplanting Religious Traditions - Asian Indians in America* (New York: Praeger, 1988).

Remembering and the Politics of Recognition

This chapter examined the transformations in the discussion of the Oriental nature of the Israeli built environment, supporting the claim that identity – as a political and cultural construct – is related to the formulation of new time and space created by communal imagination processes that intertwine past, present and future. This process is a manifestation of hegemonic culture, which frames the place while intervening and generating spatial transformation, using space production as an instrument for their realization. Thus is formed the pedagogic landscape, the spatial fabric which teaches us about our past and our identity, and within which the built environment assumes its structured symbolic significance, being justified as a representative of the collective desire and thought.

The discussion points out the role of the built environment in the production of Jewish place in the old-new space, and, as indicated, this is the site of ongoing struggles, in which top-down power creates counter-reactions that do not adhere to the desire to modernize/Westernize space. Indeed, within the Israeli space, as a product of counter-acts, the modernist housing machine became a hybrid twice. First, it became transformed from a site that expressed a unified modernist national identity into a site that symbolized the excluded Mizrahi population. Second, it became transformed from a site based on the notion of progressive modernity into a site of alternative modernity, where individual place production violates the top-down initiatives.

Indeed, space is not a static container of social relations; people create alternative local narratives that do not necessarily reflect the rationale of the nation or of capital, nor the social hierarchy or the power relations that create them.[51] In relation to this chapter, the counter-production of urban order, I would propose, is a direct result of the Zionist ideology based on the denial of a diaspora past. However, this conclusion does not aim to idealize or essentialize the Arab character of Jews in the diaspora. Following Raz-Karkotzkin, historically there had been tension among the Jews as a minority in their Arab countries of origin. However, this tension was not defined as a cultural gap to be overcome. It is the Eurocentric model of the Israeli national project that contains the East-West dichotomy as an objective category of modernity and space ordering that leads to the conclusion that being included in the Israeli collective is both the tangible and symbolic act of Jews foregoing Arab culture and place construction.[52]

Theoretically, I have indicated the relevance of postcolonial theory to the understanding of the Israeli space/place production.[53] First, the postcolonial body of knowledge critically examines the social structures that result from ideologies

---

51   James Holston, *The Modernist City – An Anthropological Critique of Brasilia* (Chicago and London: University of Chicago Press, 1989), 31–34.

52   Amnon Raz-Karkotzkin, "Exile within Sovereignty", 126.

53   In the scope of this chapter, I did not address the relevance of postcolonial discussion to the understanding of social and political structures in Israel (e.g., in *Theory and Criticism*, 20 and 26). However, one topic that postcolonial discussion has not included

of domination stemming from colonial histories.[54] Second, postcolonial criticism has enabled an analysis of the ways in which subaltern cultures are shaped while internalizing hegemonic culture. Third, a significant issue in the postcolonial theory is hybridism which is linked to the notion of diaspora. Here, allow me to rephrase the notion of diaspora as it is exposed by the case study. The concept of diaspora indeed incorporates the transnational experience of those who returned home according to Zionist ideology, a situation producing negotiable multidirectional ideas and spatio-cultural urban topographies.

Through these lenses, this chapter rethinks the traditional view of architecture that assumes the national category as the natural realm for space production – a perspective that lends high priority to official planning and architecture practice as an apparatus for the production of national sovereignty, an issue discussed in the first section. Yet, it also suggests that the multiple loyalties of people are simultaneously moulded in different spaces/places, locating themselves between here and there, within sovereign, state boundaries, and at the same time in their diaspora experience that produces a hybrid place. This term is not a fixed topographical site of negotiation between different locales (i.e., societies, cultures), but rather a zone of de-territorialization which in turn produces identity.

Hybridity as a site of negotiation was not confined merely to the tenement housing blocks environment. The new social, economic and political structure as indicated in Netivot enabled the shifting of the excluded imagined place and desires to be included in the official mapping of the city and the extensive infiltration of this architecture into other spheres became visible. In fact, the modernist model that seeks to level the range of identities has become an indication of a multicultural option that grows from the bottom up and enables a discussion of Netivot as a project of alternative modernity – a concept focusing on the significance of modernity in daily life among societies and spaces that are not part of the "first world". This type of modernity rejects the bourgeois ethos of modernity, and, instead, seeks recognition of the fact that different modernization projects have not produced uniform results.[55] At the basis of this cultural theory lies recognition for the many expressions of modernity. The capitalist economy, technology and bureaucratic organization of the state are inherent elements of modernity, but they lead to different types of modernity that diverge from the binary view of modernity versus traditionalism.

In this context, the peripheral city of Netivot can be seen as a third space, where the recognition of planning authorities enabled the expression of communal architecture that is not subjected to the hegemonic narrative. A similar argument is

---

is the position of space designing practices, such as urban planning and architecture, which challenges this chapter.

54   Jane M. Jacobs, *Edge of Empire* (London and New York: Routledge, 1996).

55   Arjun Appadurai, *Modernity at Large* (Minneapolis: University of Minnesota Press, 1996).

presented by Ben Ari and Bilu[56] who suggest that the emergence of sacred sites of Jewish saints in Israeli development towns is not just rooted among diaspora North Africans, but rather it strengthens people's sense of belonging to their places:

> By constructing these sites people in development towns come to terms with their peripheral status in Israel. This phenomenon is related to what maybe termed an internal Israeli cultural debate centering on its identity as a "Middle Eastern" society; to the extent which Israel shares with its Arab neighbors a set of cultural concepts and guidelines by which public life is carried out.[57]

The question that remains is to what extent it is possible to view the third space as an arena of subversive struggle and negotiation, as suggested by Homi Bhabha. Let me suggest that although the third space can be seen as an element that challenges the hegemonic perception of space, it does not transform it strategically. If we would return to the peripheral characteristics of Netivot (in terms of socio-economic, class and ethnic stratification that I have cited above),[58] this recognition cannot come in place of or be separated from distributive questions.[59] Rather, it should not draw attention away from distributive issues, as then the city would fall into the trap of perpetuating the hierarchy as dictated by the state's spatial ordering.

References

AlSayyad, Nezar, "Hybrid Culture/Hybrid Urbanism: Pandora's Box of the 'Third Place'". In *Hybrid Urbanism: On the Identity Discourse and the Built Environment*, edited by AlSayyad, Nezar, 1–20. Westport: Praeger, 2001.

Appadurai, Arjun, *Modernity at Large*. Minneapolis: University of Minnesota Press, 1996.

Ben-Ari, Eyal and Bilu, Yoram, "Saint's Sanctuaries in Israeli Development Towns". In *Grasping Land: Space and Place in Contemporary Israeli Discourse and Experience*, edited by Ben-Ari, Eyal and Bilu, Yoram. Albany: State University of New York Press, 1997.

---

56 Eyal Ben-Ari and Yoram Bilu, "Saint's Sanctuaries in Israeli Development Towns", 61.

57 Eyal Ben-Ari and Yoram Bilu, "Saint's Sanctuaries in Israeli Development Towns", 61.

58 New research findings point to the way in which Netivot's image among the Israeli public has been improved. One explanation for this is the transformation of the city into a spiritual node that exposes it to the public. Furthermore, in comparison to other development towns, Netivot's economy is improving, independent of state subsidies or intiatives (*Ha'aretz*, 26 January 2007).

59 Nancy Fraser, "Social Justice in the Age of Identity Politics: Redistribution, Recognition, and Participation", in Nancy Fraser and Axel Honnet (eds) *Redistribution or Recognition? A Political-Philosophical Exchange* (London: Verso, 2003).

Beng-Lan, Goh, *Modern Dreams: An Inquiry into Power, Cultural Production and the Cityscape in Contemporary Urban Penang, Malaysia*. Ithaca, NY: Cornell, Southeast Asia Program, 2002.

Bhabha, Homi, "The Third Space: Interview with Homi Bhabha". In *Identity, Community, Culture, Difference*, edited by Rutherford, Jonathan, 207–222. London: Lawrence and Wishart, 1990.

Bhabha, Homi, *The Location of Culture*. London and New York: Routledge, 1994.

Bilu, Yoram, and Ben-Ari, Eyal, "The Making of Modern Saints: Manufactured Charisma and the Abu-Hatseiras of Israel". *American Ethnologist*, 19 (4) (1992): 29–44.

Castells, Manuel, *City, Class and Power*. London: Macmillan, 1978.

Central Bureau of Statistics. *Characterization and Ranking of Local Authorities According to the Population's Socio-Economic Level in 2001*. Jerusalem: Central Bureau of Statistics, 2004.

Clifford, James, *Routes: Travel & Translation in the Late Twentieth Century*. Cambridge, MA: Harvard University Press, 1997.

Efrat, Zvi, *The Israeli Project: Building and Architecture 1948–1973*. Tel Aviv Museum of Art, 2004 [Hebrew].

Fenton, John, *Transplanting Religious Traditions – Asian Indians in America*. New York: Praeger, 1988.

Fraser, Nancy, "Social Justice in the Age of Identity Politics: Redistribution, Recognition, and Participation". In *Redistribution or Recognition? A Political-Philosophical Exchange*, edited by Fraser, Nancy and Honnet, Axel, 7–109. London: Verso, 2003.

Holston, James, *The Modernist City - An Anthropological Critique of Brasilia*. Chicago and London: University of Chicago Press, 1989.

Jacobs, Jane M., *Edge of Empire*. London and New York: Routledge, 1996.

Kallus, Rachel and Law-Yone, Hubert, "National Home/Personal Home: Public Housing and the Shaping of National Space in Israel". *European Planning Studies*, 10 (6) (2002): 765–779.

Kosansky, Oren, "Reading Jewish Fez: On the Cultural Identity of a Moroccan City". *The Journal of International Institutes* (8.3.2001), www.umich.edu/~iinet/journal/regional.

Kosansky, Oren, *All Dear Unto God: Saints, Pilgrimage, and Textual Practice in Jewish Morocco*. Ph.D. dissertation, University of Michigan, 2003.

Lefebvre, Henri, *The Production of Space*. Oxford, UK and Cambridge, MA: Blackwell, 1991.

Levy, André, "Center and Diaspora: Jews in Late-twentieth-century Morocco". *City & Society*, 13 (2) (2001): 245–270.

Madanipour, Ali, *Design of Urban Space – An Inquiry into a Socio-Spatial Process*. Chichester: John Wiley and Sons, 1996.

Nitzan-Shiftan, Alona, "Whitened Houses". *Theory and Criticism*, 16 (2000): 227–232 [Hebrew].

Norberg-Schultz, Christian, *Genius Loci – Towards a Phenomenology of Architecture*. New York: Rizzoli, 1979.

Oz, Amos, *In the Land of Israel*. Tel Aviv: Am Oved, 1983.

Posner, Yaakov, "The Village in Eretz Israel". *HaBinyan* (1938): 1–2 [Hebrew].

Raz-Karkotzkin, Amnon, "Exile within Sovereignty: Towards a Critic of the 'Negation of Exile' in Israel Culture". *Theory and Criticism*, 4 (1993): 23–56 [Hebrew].

Relph, Edward, *Place and Placelessness*. London: Pion, 1976.

Safran, William, "The Jewish Diaspora in a Comparative and Theoretical Perspective". *Israel Studies*, 10 (1) (2005): 36–60, http://muse.jhu.edu/about/publishers/indiana.

Schiler, Abraham, "Land Development Problems for Housing". *HaBinyan* (1937): 28–33 [Hebrew].

Shadar, Hadas, "Between East and West: Immigrants, Critical Regionalism and Public Housing". *The Journal of Architecture*, 9 (2004): 23–48.

Sharon, Arie, *Physical Planning in Israel*. Jerusalem, 1951 [Hebrew].

Sharon, Smadar, *To Built and be Built: Planning the National Space in the Formative Years of Israel*. Thesis submitted at Tel-Aviv University (2004) [Hebrew].

Shenhav, Yehuda, *The Arab Jews: A Postcolonial Reading of Nationalism, Religion, and Ethnicity*. Stanford: Stanford University Press, 2006.

Taylor, Charles, "Two Theories of Modernity". *Public Culture*, 11 (1) (1999): 153–174.

Tuan, Yi-Fu, *Space and Place: The Perspective of Existence*. Minneapolis: Minneapolis University Press, 1977.

Tzfadia, Erez and Yiftachel, Oren, "Between Urban and National: Political Mobilization among Mizrahim in Israel's 'Development Towns'". *Cities*, 21 (1) (2004): 41–55.

Yacobi, Haim, "The Third Place: Architecture, Nationalism and Postcolonialism". *Theory and Critique* (2007) [Hebrew].

Yacobi, Haim, "From State-Imposed Urban Planning to Israeli Diasporic Place: The Case of Netivot and the Grave of Baba Sali". In *Jewish Topographies: Visions of Space, Tradition and Place*, edited by Baruch, Julia, Lipphardt, Anna and Nocke, Alexandra, 63–82. Aldershot: Ashgate, 2008.

Zaslevsky, David, *Housing for Immigrants – Construction, Planning and Development*. Tel Aviv, 1954 [Hebrew].

Zaslevsky, David, *A Survey of Netivot's Development*. Jerusalem: Ministry of Housing, 1969 [Hebrew].

Chapter 3

Neighbourhood and Belonging: Turkish Immigrant Women Constructing the Everyday Public Space

Eda Ünlü-Yücesoy

Neighbourhood is addressed ideally in the urban studies and planning literature with qualities of community building, social cohesion and, particularly where migrants are concerned, the place of social integration. An important spatial unit for all inhabitants, starting from the individual housing unit and extending to the city, the neighbourhood stands in the medium of multiple dimensions associated with the private and public realms. Though designated as the social ground on which inhabitants are allowed to feel part of their community and build social contacts, the neighbourhood public space is conceived to a large extent by its physical/configuration characteristics rather than as a social construct. While physical perception refers to an enclosure, i.e. what constitutes the public space and how its framing elements determine the level of use of the public space, and a container, i.e. the type, size, and kind of activities and how land-use patterns affect the state of the public space; the predominating land-use activity (commercial, residential, recreational); the mode of its use, whether it is a place for staying in (activity space) or a channel of movement for people or vehicles; the latter, public space, as a social construct, calls for recognition of the neighbourhood public space as an interactional and experiential space, defined and mediated by social conventions, rules and regulations, and symbolic boundaries. In this framework, users are social actors engaged in the continuous process of place-making, exploring, negotiating and appropriating the public space. Experiencing the public space, then, becomes not a simple experience of a physical setting, but a distinctive social reality. It is on this ground that this chapter sets out to examine the relationship between the everyday neighbourhood spatial practices of Turkish immigrant women, with regard to the social construction of urban public spaces and formation of everyday belonging, and representations of neighbourhood public space in the Dutch context. The diversity of Turkish immigrant women's social constructions of neighbourhood urban public spaces, which in turn shape or frame their spatial behaviour, are elaborated in relation to different characteristics and kinds of neighbourhood spaces, their planning and design.

Employing the neighbourhood public space as a social construct calls for accommodating a multivalent representation of space, how it is constructed and experienced as material artefacts and how it is represented, i.e. 'spatial practices' and 'representations of space' (Lefebvre, 1991). Representations of space present the dominant (imposed) spatial order on the rhythms and rituals of everyday life. Spatial practices reflect different, sometimes contested and conflicted, constructions and at the same time possibilities/restrictions of the further appropriation of public spaces. Whilst representations operate abstractly for making the professional codes compelling for decision-making, the overall spatial practices that people perform and evaluate in and about a particular space set norm assumptions about that place (Modan, 2007). Thus, spatial practices can be congruent with or challenge the representations of space, yet they persist. In this framework, analysis of the dialectical interplay between the representations of space and spatial practices has a revelatory importance in elucidating the relations between space in use and identity in process. As Turkish immigrant women's spatial practices are deciphered in their conceptualization and experiences of neighbourhood spaces, where their everyday social relationships are formed, and how they construct belonging and dis-belonging in the neighbourhood are revealed. According to Fenster (2004), everyday belonging can mean to be rightly placed or classified or to fit in a specific environment, or something associated with past and present experiences and memories and future ties connected to a place. In that sense, everyday belonging is closed to 'habitus' (Bourdieu, 1986), a sense of one's place, an embodied sense of place. Referring to the "embodiment of individual actors of systems of social norms, understandings and patterns of behavior",[1] habitus reflects personal accumulated space-time experiences and inheritances. It is not only rules. The social norms and values that habitus represents enables us to use the knowledge coming from past experiences to direct spatial practice, but also improvisations that people develop throughout their lives as a 'sense', a sense of how to act. If habitus is as an objective basis for regular modes of behaviour and also for the regularity of modes of practice, certain meanings and experiences attached to spaces are then important for the realization of practices. As habitus implies a 'sense of one's place', at the same time, it constructs categorical identifications, 'a sense of the place of others'. Based on the similarities of habitus, a group may act similarly and, in turn, reproduce the culture of their shared social fields through practice (Bourdieu, 1986). In this way, we identify ourselves and others, classifying and categorizing 'we' and 'they'. Thus, habitus accommodates the articulation of symbolic and social boundaries, working in categorization of social and collective identifications and underlying the difference between 'I' and 'you', and 'we' and 'they'. In the public spaces, these symbolic and social boundaries are always at play among individuals and groups, constant in competing, struggling, negotiating for the articulation of differences and communalities and patterns of appropriation of public spaces. In this framework, different forms and varying degrees of belonging and dis-belonging, such as avoidance and participation, withdrawal and placement, are articulated in the relational construction of everyday public spaces.

---

1   Painter, 2000, cited in Hillier and Rooksby (2005: 21).

Representations of Neighbourhood Public Space in the Dutch Context

The research from which this chapter is drawn is based on a contextual and exploratory analytical study[2] on everyday public spaces in two neighbourhoods of Enschede, a middle-sized city in the east of the Netherlands. Deppenbroek and Wesselerbrink are typical examples of the modernist functional city planning[3] and working-class suburbs that came into existence with a certain vogue in architectural and social design and dominated the public housing provision in many Western European cities. In addition, the romantic/intuitive approach of the Delft School was influential in Enschede. Favouring the image of village in contrast to the city, neighbourhoods were seen as 'pillars'[4] that function as smaller and more natural communities. Though entire 'pillarized' neighbourhoods were never realized, the idea prevailed in design that the neighbourhood comprised of a particular pillar would function as a unique cohesive community.

Deppenbroek was planned in the early 1950s with the principle of 'sun, space, and greenery', with residential areas and rows of buildings built in semi-

---

2   Forty-seven Turkish immigrant women were interviewed regarding their use and experience of various spatial scales of urban public spaces in the city. Those Turkish women, whose presence in the Netherlands is for a definite period of time and who have limited-time residence permits, like student and visitor visas – even if it exceeds a six-month period – were not included in the fieldwork. In addition, because of political concerns, there are no respondents from the various ethnic groups of Turkey, such as Kurds, Arabs and Christians. Among the Turkish respondents are different religious affiliations and brotherhood organizations, such as Alevis and Sunnis. All of the interviews were conducted in Turkish and Dutch. English translations were provided by the author. One of the key informants, who is of half-Dutch half-Turkish origin and works as an advisor on immigrant issues affecting Turkish inhabitants in the Overijssel Province, helped the author translate some words from Dutch to Turkish and Dutch to English. For the sake of precision, the translations are not interpreted in order to give true accounts of the respondents' real experiences. In that sense, while quotations are made from interviews, some Turkish and Dutch words are kept in the original in order to preserve minutiae. Italics in parentheses in quotations are the authors' additions in order to give a good grasp of the quotation. The age and permanence in the Netherlands of each respondents are stated at the end of each quotation.

3   Although the work of Corbusier started to be highly influential in Europe and the Netherlands in the 1920s (following the declaration/publication of Le Corbusier's *The Contemporary City for Three Million People* in 1922 and meetings/congresses of the CIAM), in most of the pre-war functionalist urban design idea(l)s of the open and functional city, his concepts of 'towers in a park', and 'a balanced community' gained popularity in the post-war period.

4   Pillarization can be described as a peculiar Dutch approach involving the mutual presence of several religious, linguistic and social cleavages. Being largely effective before the Second World War and in the 1950s, 'pillarized' society was based on religion and ideology as central social determinants, shaping every aspect of social life. Pillars operated separately from each other and established their own social institutions, such as their own places of worship, schools, hospitals, sports clubs, political parties and newspapers.

closed or open block formations, a mixture of hallway-access flats or high-rise gallery-access flats with single-family row houses, strict traffic regulation with connections to the city centre and other neighbourhoods, and spacious green areas (see Figures 3.1 and 3.2). Wesselerbrink, on the other hand, was planned in the late 1960s with the ideals of *wijkgedachte*, integral neighbourhood unit planning, and *wooneenheden*,[5] designed on the basis of the concept of the *brink*[6] (see Figure 3.3). Criticizing the inflexible and mechanical concepts of orderly planning, the neglect of public space, and the lack of relationship between the buildings and their surrounding context, *wijkgedachte* was motivated by the idea of a city organized as a series of interlocking social circles starting with the family home and extending up to the neighbourhood to the city as a whole, reflecting a constant interplay between the public and private domains within linear mass-produced buildings which facilitated individual mobility.[7] While the social basis was composed of the promotion of neighbourliness and face-to-face interaction among people living in the same neighbourhood, the physical basis was to create the necessary built environment enabling the desired social interaction and creation of community in the neighbourhood. The idea of the community was again place-based.

In the late 1970s, most of these working-class neighbourhoods had a dynamic mixture of different groups and high rates of interchanging tenants. While social climbers left these neighbourhoods to live in the suburban areas in spacious housing units with their families, the elderly stayed and students and guest workers' families started to replace the leavers. As the number of leavers increased, so did the number of guest workers' families residing in the neighbourhood from the 1980s onwards. Although there are no examples of immigrant ghettos literally in the Netherlands, the concentration of low-income social housing in certain neighbourhoods enables the creation of pronounced immigrant clusters. In 2000, the 'strategic district approach' was initiated in neighbourhoods with a uniform housing stock and a correspondingly skewed population distribution. Labelled as the incubators of multiple deprivation and high social fragmentation, these post-war neighbourhoods are subjects of nationwide urban renewal plans and programmes.

---

5   *Wooneenheid* is an ensemble of housing units within the neighbourhood and, in social and city planning terms, an iterable unit, composed of different housing forms, such as low-rise, row-houses, or housing for old people, for households in different demographic characteristics, e.g. old, single, and families with children.

6   The *brink*, according to Van Dale dictionary, has two meanings. The first one is a grass-yielding yard around a farmer's house, and the second is the name of the village square in the east Netherlands, a flat area often circled with trees with a church in the middle. The design of *brinks* in Wesselerbrink is based on the premise that every unit/cluster/community has its own centre and, depending on the scale of the unit/cluster/community, there are facilities for shopping, education or green units for the benefit of the inhabitants of the unit/cluster community concerned.

7   Mumford (2002: 5–13).

Figure 3.1   Deppenbroek, Enschede

Figure 3.2   Wesselerbrink, Enschede

Figure 3.3   Wesselerbrink, Enschede

Exploring Turkish Immigrant Women's Appropriation of Neighbourhood Public Spaces

Generally described as group-oriented, introvert and bound to religious/ethnic norms, Turks in the Netherlands give great importance to the neighbourhood in which a house is located, and favour cultural similarities among neighbours as well as the presence of Turkish elements.[8] As the single-family dwelling is the most desired form of housing, social characteristics are pronounced strongly, too.

> I have lived in this house for 28 years, and last year the company decided to sell it and luckily we were able to buy it. In time, my sisters have also moved here, with one of them, there are just four houses between us. When my son got married, he found a house on a back street. Now he also has opened a business in the shopping centre. My life is here. I have never thought of leaving this house or moving to another neighbourhood. All of my family is here. What would I do somewhere else? Maybe the house could be better, or much comfortable, but no, no, we are here together. (b. 1955, 34 years in NL)

---

8   SmartAgent (2001: 47).

> We bought this house last year, but I cannot get used to living in this neighbourhood. I grew up in Deppenbroek and all my family, friends, everybody are there. When I go there, I feel at home. Not necessarily seeing my family or friends, every face in the street tells me that I know that person, well, in fact not, but it feels like that ... *thuis voelen* (feeling at home/at ease), do you understand? I wish we could find a house there in Deppenbroek. (b. 1974, b. in NL)

The high preference for the single-family dwelling in the accessible housing stock increases the likelihood of Turkish families agglomerating in certain neighbourhoods. Voluntary concentration fosters the construction of several Turkish communities and community associations, sharing more or less similar habitus.

Home reflects the identity of the inhabitants through a variety of habitation manners and styles. Through open or closed curtains, satellite dishes, various ethnic ornaments and decorations, the identity of the inhabitants are exposed.[9] For the passerby in the street, ethnicity is in open display; while Dutch residents keep their curtains open, day and night, Moroccans keep theirs closed all the time, and Turks prefer them half-opened, half-closed. Home as the centre of the private domain thus opens itself to the public. Apart from these passive exhibitions, there also can be intense identity displays in the home cluster, particularly in the street. The contradictions among users' conceptualizations of publicness and privateness become visible in these spaces.

> The expanding Turkish community started to manifest itself more openly and self-consciously in public. On sunny days the women and children would group together on street corners and green areas. The men repaired their cars on the street. Neither of these activities was in line with the privatized living culture of the established residents ... The combined 'openness' and 'privateness' of 'modern' Dutch participation culture ... came into conflict with the closed and collective character of 'traditional' ethnic minority cultures. (Mommaas, 1996: 206)

The importance of the privacy of the Muslim home reflects itself not only in the architectural characteristics of the house, but also the social characteristics attributed to the house itself. Rapoport (1977) makes another important distinction in terms of the conception of privacy, that is, 'the privacy is for the group, not the individual'. The home cluster, as the space for a particular group, therefore gains importance to socialize and sustain the daily relations. The home cluster comprises much of the neighbourhood use for Turkish women. They find it significant how the home is seen outside, both in the eyes of the Dutch and other migrant groups and the Turkish community in the neighbourhood. The space in front of the home demonstrates decency in order to gain recognition and respect from neighbours. It should be clean, neat and modest. Where the house meets the street, the 'outside' is transformed into the 'inside'. In both neighbourhoods, single-family houses on

---

9   Horst and Messing (2003).

curbed streets or cul-de-sacs welcome broad use of the front, the semi-private/ semi-public space for gathering, working their handcrafts, drinking tea, and talking, while the children play in the street. The space between the home and the street becomes a social space not only for the residents, but also for people passing by, facilitating greetings and short conversations among neighbours.

> Sometimes we do place the chairs just in front of the house, me and my neighbours. We sit and talk. When the weather is good, it's nicer to be out than inside. Then you see people around.
>
> *Do you also use this place with your family, for dinners or family visits?*
>
> No, no, it's just us women. Only neighbours, not friends. (b. 1969, 15 years in NL)

The claim on the home is quite strong in a sense that it expands to the home cluster. Here, they openly express what is the appropriate behaviour and what is not. In this way, the moral geography of the neighbourhood is constructed. The claims over space also set the social and cultural codes and norms for the general Dutch society.

> You never see a normal decent Dutch person sitting in front of their houses and drinking alcohol, also no Turkish men do that … They are only those Gypsies [referring to the lower class Dutch men as 'gypsies'] sitting in front of the house and drinking beers. (b. 1960, 20 years in NL)

Home expands further to parks and green areas, by 'mobile home territories'.[10] The engagement of 'travelling packs' and the establishment of 'mobile home territories' in public is strategic, especially for first-generation Turkish women whose networks are limited to peers. Though for Lofland (1973), travelling packs are another method of avoiding the city while living in it, for Turkish women it is an important strategy of exploring the city. With travelling packs and the creation of home territories, they eliminate any need for concern about establishing their identities with strangers, as well as discovering the presence of strangers among them. In addition, they can reduce the negative concerns in the community as the acknowledged 'with' persons provide reassurance and support when necessary. As a group, they can avoid unwanted contact. Through the repetitive use of mobile 'home territories', their spatial knowledge grows and, as in the case of many marriage-immigrant women, they start to get used to 'how it is to be in public'. Though neighbourhood parks and playgrounds comprise the home territories of the Turkish women from all generations, the frequency of use, the size of the groups, and the duration of stays are different. The creation of home territories through the appropriation of parks by older women also acts as a control mechanism that is sometimes quite intense for the young and unmarried.

---

10   Lofland (1973: 137).

Not all public spaces provide promise for the creation of home territories. Suitable places are learned and tested on every visit; therefore suitability is granted with extensive local knowledge and increasing familiarity. They ensure appropriate times and patterns of uses of others.

> We usually go to the park during the long summer evenings, when there's still light. At that time, there are also other families walking and entertaining their children. Also, at that time, people who let their dogs out in the park are few. During the day, it's nice, but the old Dutch people usually come in the morning and afternoon; later, teenagers come in groups. For us, late afternoon, say around six, is quite good to go there. (b. 1951, 26 years in NL)

In addition to the parks, the *volkstuin* or *moestuin*, small-scale agricultural places leased on the outskirts of the city for growing vegetables and plants, exhibit home-like characteristics in terms of the relations and meanings attached. Providing seasonal vegetables for the household, the *moestuin* enables them to have a social place, a meeting ground for family members and friends, and to maintain a rural atmosphere in the city. Especially in the summer, they spend the day at their *volkstuin*, working with the plants, hosting their visitors, and visiting their friends' *volkstuinen*.

> When I first came here, my neighbours [Turkish, my addition] were telling me nice stories about their *moestuin* and how they had enjoyment with the other gardeners … having lunches altogether, tea afternoons. They went there together and they spent the whole day together. This year I also rented one. (b. 1967, 19 years in NL)

The most important feature of the use of urban public spaces around the home and those exhibiting home characteristics in terms of social relations is that these spaces are strictly gendered. Not only religious women, but also secular women mention an 'understanding' in conducting activities with other women. Men usually spend time outside the home or neighbourhood, in a specific place for men, like a coffeehouse, or they stay inside the house. The presence of men in front of the house is considered inappropriate, breaking the codes of deference and modesty. The home cluster thus becomes the place of construction, identification and expression of 'we'.

Conceiving the Neighbourhood: Are Turkish Women 'Urban Villagers'?

The majority of Turkish women in Enschede are of rural origin, directly implanted in the Dutch urban life without any previous urban experience. This has had profound effects on their conceptions and interpretations of the general urban area and neighbourhood. Gans (1982) talks about the urban Italian American villagers in Boston's West End. Similarly, Lofland (1973) describes an 'urban village' as a neighbourhood, "a home territory writ large". An urban village roughly corresponds with living the personal world in the midst of urban anonymity. Ethnicity is an important binding element, yet not the sufficient one.

All of the first-generation respondents have lived in the same neighbourhood since they came to Enschede. Among the Turkish immigrants, *hemşehrilik*[11] is an influential institution and a powerful tie. Because of the dispersed geography of Turkey, every village has its own cultural, social and religious traits, therefore coming from, for example, a town in the eastern Black Sea region implies traditions, habits and customs very different from those of a town in the western or central regions. In the migration process, the selection of guest workers from the same regions, even from the same village, has helped the Turkish community to form, if not keep totally, their social networks in the line of *hemşehrilik*. For the Turkish women, therefore, who arrived a decade after their husbands in the Netherlands, the first social networks ran parallel to those of their husbands' and were based on the *hemşehrilik*. Though, as they settled, they might build their own social relations with other Turkish women, most *hemşehrilik* relations continued. In addition, since most Turks can only afford 'social housing' in certain neighbourhoods, they have found housing in the same neighbourhood. *Hemşehrilik* has not perished, on the contrary, it has strengthened even more with time from living in the same neighbourhood.

The urban villagers' entire pattern of daily life passes within the limits of the neighbourhood or even more narrowly, within the limits of personal relations. For the first-generation and marriage-immigrants, the neighbourhood corresponds to their whole life in Enschede.

> Enschede is Deppenbroek for me. Here you can find everything. There is no need to go to the centre. Well, if you are very sick, you need to go to the hospital, but here we have a doctor, dentist, everything, in the shops. What more do you want? Clothes, food, everything. (b. 1953, 23 years in NL)

> I usually go out in the neighbourhood to take a walk with my children. They are bored at home, and we walk together in the streets, to the shopping mall, to the market. My husband works hard, and on the weekend either he rests or he gets together with his friends. With the children it's very difficult to go to the city, without a car, impossible, and also I don't need to, we have everything here in Wesselerbrink. (b. 1978, 23 years in NL)

Together with *hemşehri* ties, limited spatial mobility and the abundance of all necessary services in the neighbourhood pave the way to the creation of 'urban villages'. In addition, since the first-generation and marriage-immigrants tend to form their social relations with peers, the encapsulated networks help them to conceive the neighbourhood as a sufficient spatial unit. Most likely, this is why the level of satisfaction with the neighbourhood and housing area is high among the

---

11   *Hemşehrilik* means 'fellow townsmanship'.

respondents. They leave the neighbourhood once a year for a trip to the city centre or to another city, for emergencies or to visit family or relatives, with the help of the family members and friends.

Neighbourhoods with high Turkish immigrant concentrations provide Turkish women, especially peers, with social climates very similar to those of their places of origin in Turkey. Because of their previous experiences of strong neighbourliness, they give special significance to the feelings, neighbourly relations and the identity of neighbours.

> When I first came to Enschede, I lived two years in Meeuwenstraat and there we had very good relations with our neighbours [Turkish]. When I moved to Poterstraat, my neighbours had also moved here, to the same street; wherever one neighbour goes, the others follow. Now we are apart [because of the firework disaster, they were placed on another street in the same neighbourhood], but none of us has gone far. I am very happy here, I have never thought about leaving this neighbourhood. We will all come together again. (b. 1953, 23 years in NL)

Conceiving the neighbourhood as an urban village has important consequences for the social construction of the public and private. In neighbourhoods, where Turkish enclaves[12] agglomerate, everyone is 'a significant other'. When they are out, Turkish women are visible everywhere and all of their actions are watched.

> I think if I lived in Deppenbroek, I wouldn't do many of the things that I do now. There everybody knows each other; it's like Turkey. What you do every day is known and discussed: why do you go to one shop, but not the other, why you go out, etc., it's like they keep records of you. (b. 1979, b. in NL)

For those who want to keep their traditions and customs, these neighbourhoods are the perfect places to continue the habitual-traditional social environment. The patriarchal lines continue and the newcomers or offspring are not alienated from the traditional values and norms, which are represented, continued and strengthened by the elderly. The ethnic networks, especially the kin-related ones, which originally functioned as support mechanisms, prevent offspring from testing alternative lifestyles in the Dutch context and perpetuate the traditional norms of gender interaction. The presence of women in public easily can be scrutinized as to whether it is intruding on Turkish traditions, customs and values.

---

12   Although there are no data available to trace Turkish enclaves in the neighbourhoods of Enschede, or for other Dutch cities, from the interviews it seems that there are clusters of Turkish groups that are quite differentiated in terms of place of origin, religious affairs and orientations. For example, the respondents pointed out specific locations in neighbourhoods as distinctive social worlds and images; the community of a religious sect with its own secret mosque, the street where women suffer excessive social control, the cluster of inhabitants from a same province in Turkey.

Everyday Routines and Repetitions of Daily Activities in the Neighbourhood

Social relations in repetitive activities and daily-life routines, located in time-space, can have regularized consequences in the boundary-making processes for individuals who engage in those activities. Everyday routines install Turkish women's spatial literacy in the neighbourhood. The neighbourhood is marked on an area of routines that relate closely to the role inventories and activities, such as frequented pathways, playgrounds and shopping malls. Mothers go shopping in the morning; they meet with other mothers in the schoolyard, while waiting for their small children to come out for lunch. Then they return them to school for another two hours. After school, they walk home with them, if the weather is suitable, they take a trip to the park. Then they walk home again. In the neighbourhood, they use the same streets probably several times a day, so that they know everything, who lives in that house, whose car that is, the old lady was not sitting in the usual place today, is there anything wrong, they broke the glass wall of the bus-stop again. Routines provide a sense of place and a sense of control and at the same time a sense of belonging and attachment.

Knowledge or sense of place arise from routines, and helps them develop familiarity both for the places and for the other users. In this sense, the anonymity of the public spaces has been broken or at least decreased. They decode the public space, testing their presence in public all the time, as well as struggling with the negative territorial concerns.

> I know very well when I can walk the sidewalks of the park. There stands in the
> late afternoon usually the teenagers, intimidating foreigners all the time. So, you
> know when to pass, you don't let your nerves down. (b. 1960, 20 years in NL)

Through knowing the places, Turkish women are able to develop 'tactics' (de Certeau, 1984) to use. Each public space has its own proper times when it can be appropriated. Similar to the use of the neighbourhood park, as discussed above, the shopping centre has its 'social times', day time on the weekdays, especially in the early afternoon hours 'when it is quiet'. In this framework, long permanence in the neighbourhood enables them to form and sustain clusters of rather homogeneous interaction systems. By appropriating space for themselves, their lifestyles and identities become more visible and perhaps non-problematic, as in the case of peculiar residential and public uses. In this way, they do not come into contact with a different culture, social habits or norms. Neither washing cars in the street nor picnics in the neighbourhood park are disputed among neighbourhood inhabitants. The social mechanisms of bonding, especially social control, can then be practiced easily.

> This is a neighbourhood where everyone looks after each other and cares for
> each other. For example, if I see a Turkish boy loitering, I say right away, 'don't
> you have better things to do?' and if I see him, for example, smoking a cigarette,

then I immediately tell his mother … and I know that other people here [Turkish people] would do the same, if it were one of my children as well.

So, you think other people may not likely take such an action?

No, no, it is what we are, we care for each other. (b.1953, 23 years in NL)

Strict functional separation, especially mono-functional housing clusters enables the creation of enclaves, socially and culturally stable units. For upwardly mobile Turkish women, the presence of these clusters in the neighbourhood is often considered an obstacle.

Those who live on some streets in Deppenbroek, in the ones looking at each other, for example in Ganzediepstraat, see inside of the each other's homes, so that you have to be always careful. (b. 1979, b. in NL)

I never let my children play outside and we have almost no relations with the Turks here in Deppenbroek. They are at the same point as when they came to the Netherlands 20 years ago. Yes, we live in this neighbourhood only because we found the best house we could afford. (b. 1968, 32 years in NL)

Rigid functional separation accommodates territorial claims easily. Playgrounds and green areas easily become places of disorder, as they are only frequented by gangs and youth groups, or become territorialized by other ethnic groups.

[showing the small playground just in front of the house] I have never seen any children playing here, only teenagers come together. They drive fast, make a lot of noise, sometimes they drink alcohol and then they leave all the garbage in the park … We usually go to other playgrounds, there are many playgrounds, so that you can choose, but a lot of them are like this. (b. 1969, 15 years in NL)

When we first moved to Wesselerbrink, we were frequent visitors of the park, sometimes with friends [neighbours] and children, sometimes with family after work. When the days are longer in the summer you can stay long outside, it's only five minutes away. When we moved here, this neighbourhood was a nice neighbourhood, few foreigners, mostly Dutch. Then they arrived, people think they are Turkish, but no, they call themselves Suryoyo's.[13] Now it's their neighbourhood, their park. (b. 1951, 26 years in NL)

---

13   Suryoyo's are generally considered to be Syrian-Orthodox Turks, who were living in south-eastern Anatolia and North Syria. Because of the political problems and instability in these regions, they immigrated to Europe during the First Gulf War, in the first half of the 1990s. Since then, problems have arisen between them and Turkish guest worker immigrants, mainly related to the separatist ethnicity concerns in Turkey.

Home cluster and home territories represent the strongest form of Turkish women's claims over the public space; from the part of casual daily communication to demarcations of legal rights, from exposure to social norms of use to specific gender practices. The 'home' represents the habitualized practices and places of conformity and manipulation. Since claims are staged openly, it is the place of interaction. While claims over public space are not as strong as those over the 'home', distant places can become, to a limited extent, an extension of the home. When the spatial literacy of the public space is achieved, there is room for appropriation, for developing tactics to use. The neighbourhood represents familiar practices and places of improvisation and contestation. In the neighbourhood, Turkish women negotiate the boundaries of public space.

Conclusion

Traditionally, neighbourhood is a very important social area in daily life for Turkish women. The neighbourhood conceived is identical with feelings of togetherness and solidarity. Even the family is embraced by the neighbourhood. Strong neighbourly feelings, not necessarily reinforced by kinship ties, are counted as very important. Neighbours where no family ties are present form a social district community just as strong. 'Buy a neighbour, not a house!' is a very common Turkish proverb. When the Turkish immigrant women's everyday spatial practices in the neighbourhood are examined in relation to the representations of public space, divergence and congruence are found in the same process.

Deppenbroek and Wesselerbrink represent typical Dutch post-war neighbourhoods, designed according to the principles of the functional city and developed according to two main goals: to care for social hygiene and health and to establish social order by means of physical planning. These goals reflect a perception of passivity of space, a taken-for-granted element as a container of buildings, dwelling units, activities, facilities and users. This static perception also can be seen in the design configuration of the relationship between the public and the private. Strict functional separation and segmented activities lead to the closure of openness and spontaneity. The lack of differentiation in housing styles, the repetitive use of apartment blocks and single-family dwellings create a passivity in space, a monotony where the low-level order hinders the astonishment of discovery. All respondents mentioned a feeling of vanishing when they first arrived. 'I couldn't go out alone, not even to the supermarket two streets away'. At first it was very difficult for her to distinguish her apartment block from the others. 'All houses and apartments were the same; yes, there are numbers on each block, but from a distance, when you walk on the street, it is very difficult to tell them apart' (b. 1967, 19 years in NL). The abolition of the street from public experience and its placement as a traffic corridor rather than a social setting hinders and restricts the spatial knowledge that can be gained unconsciously by routines and everyday activities. The perception of the street as a mere transit space at the same time interrupts the hierarchical relationship between the public and the private, leading to blurry conceptions of control in the public space. This ambiguity due to design accelerates territorial conflicts of appropriation, leading to avoidance and withdrawal, especially for those who are controlled – the weak and disadvantaged. In that sense, a 'strongly classified spatial system' (Sibley, 1995) dictates covert strategies and tactics for users. Everyday practices are shaped to a certain extent as processes of appropriation and territorialization; power relations divided between the dominant and dominated, strategies and tactics, and place and time. It is no surprise that Turkish women's appropriations rely on seized opportunities and on cleverly chosen moments, like visiting the shopping mall in the neighbourhood when 'it is quiet', in the early mornings on the weekdays, or walking in the neighbourhood or in the park just after dinner, before it gets dark and before the Dutch neighbours let out their dogs.

Ambiguities of control in the public space facilitate the creation and strengthen expansion of home territories, particularly in the clusters in which Turkish immigrants live predominantly. These home territories are strategic, especially for first-generation and marriage-immigrants, helping them to preserve their habits and customs in Dutch society. In this sense, these territories also can be seen as 'a blessing for those who tend to carry-out their traditional gender roles and rural ways of life in their new environment' (Erman, 1998). It can be further argued that the 'rurality' and 'Turkishness', i.e. traditions, customs and habits that they brought from the place of origin, as a particular way of appropriation of space, are preserved. In addition, these places help Turkish women to reproduce traditional gender roles, as well as gendered practices. On the other hand, with home territories, it is possible for Turkish women to stage themselves in the public space, revealing their own social norms of conduct. In this sense, Turkish women's home territories also can be seen as domains of exposure in which they claim the public space.

The planning ideas behind the post-Second World War neighbourhoods aimed at creating coherent communities. In that sense, from the Turkish immigrant women's spatial practices in the neighbourhood, it to a certain extent has been achieved with the creation of urban villages. This can be due to two reasons. First, the large proportions of social housing units with certain ethnic groupings enable Turkish immigrant to articulate vertical bonds of *hemşehrilik* with horizontal bonds of friendship in everyday routines and fortify the urban villages. Second, as Boomkens (1999) states, by implementing the 'sun, space, and greenery' principle instead of concentric urban growth, the intrinsic characteristics of what makes the urban and 'rural' distinctive disappears. The interweaving of green and at the same time vacant spaces with densely built housing areas blurs the line between urban areas and rural villages; the emptiness of green areas and over-planned urban areas become silent as if they are villages and the homogeneity of the monoculture in the neighbourhoods resembles the closeness of the local/agrarian culture. In that sense, Turkish women's past experiences, directing their social relations, cannot be challenged or changed due to changes in habitat, and therefore the old habitus prevails.

Turkish women's patterns of use and appropriation of everyday public space points out belongingness and dis-belongingness as expressed in their daily practices in different scales of public spaces in the neighbourhood. For Fenster (2004), a claim over public space is one of the expressions of belonging in everyday life. The potency of these claims determines the degree/volume of belonging. These degrees/volumes of belongings, their involvements and attachments, are their social practices becoming *implicated*[14] in the physical space. Their involvements and attachments, the way that they are connected to public space, reveal the meanings of these places, mechanisms of exclusion and inclusion, of retreat and interaction. They, at the same time, expose their specific senses of place, their habitus, relational frameworks of embodied dispositions.

The study of Turkish immigrant women's conceptions of neighbourhood public spaces provides planners and policy-makers with deeper knowledge on not only how these women think about, use and value their living spaces, but also, and more importantly, how they 'flourish', i.e. how they define, express,and utilize their rights and needs while creating and transforming spaces. Therefore, 'potential' (Gans, 1968) public spaces, those constructed by planners from above within certain social and cultural conditions, may become 'effective' public spaces. For planners and policy-makers, it offers new ways of thinking about urban life and how inhabitants actually live and particularly realizing the relevance of the everyday spaces of the housing cluster or neighbourhood in the daily lives of inhabitants.

References

Boomkens, René. '"Van de grote stad ging een onbestemde dreiging uit" Hoe grootstedelijk is Nederland?'. In *De stad op straat*, edited by Ries van der Wouden, 63–80. The Hague: Sociale en Culturele Studies 27, 1999.

Bourdieu, Pierre. *Distinction; a Social Critique of the Judgement of Taste*. London: Routledge and Kegan Paul, 1986.

De Certeau, Michel. *The Practice of Everyday Life*. Los Angeles: University of California Press, 1984.

Erman, Tahire. 'Semi-public/semi-private spaces in the experience of Turkish women in a squatter settlement'. *Stadtplanung*, 22, special number, 'Symposium Migration und Offentlicher Raum in Bewegung' (1998): 40–51.

Fenster, Tovi. 'Gender and the city: The different formations of belonging'. In *A Companion to Feminist Geography*, edited by L. Nelson and J. Seager, 242–256, 2004. www.tau.ac.il/~tobiws/gender_companion.pdf (accessed 19 November 2005).

Gans, Herbert J. *People and Plans: Essays on Urban Problems and Solutions*. New York: Basic Books, 1968.

Gans, Herbert J. *The Urban Villagers*. New York: The Free Press, 1982.

---

14   Gow, 1995, cited in Waterson (2005: 334).

Hillier, Jean and Emma Rooksby. 'Introduction to first edition'. In *Habitus: A Sense of Place*, edited by Jean Hillier and Emma Rooksby, 19–42. Aldershot: Ashgate, second edition, 2005.

Horst, Hilje van der and Jantine Messing. 'Bij Marokkanen zijn de gordijnen dicht'. *Agora*, 19, 4 (2003): 14–16.

Lefebvre, Henri. *The Production of Space*. Oxford: Blackwell, 1991.

Lofland, Lyn H. *A World of Strangers: Order and Action in Urban Public Space*. Illinois: Waveland Press, 1973.

Modan, Gabriela G. *Turf Wars, Discourse, Diversity, and the Politics of Place*. Malden: Blackwell, 2007.

Mommaas, Hans. 'Modernity, postmodernity and the crisis of social modernization: A case study in urban fragmentation'. *International Journal of Urban and Regional Research*, 20, 2 (1996): 196–216.

Mumford, Eric. 'From CIAM to Collage city: Postwar European urban design and American urban design education'. Paper presented at the conference, Urban Design: Practices, Pedagogies, Premises, Columbia University, USA, 5–6 April 2002. www.arch.columbia.edu/gsap-online/files/00/00/00/13099/Briefing%20Materials.pdf (accessed 18 September 2005).

Rapoport, Amos. *Human Aspects of Urban Form: Towards a Man-Environment Approach to Urban Form and Design*. New York: Pergamon Press, 1977.

Sibley, David. *Geographies of Exclusion: Society and Difference in the West*. London: Routledge, 1995.

SmartAgent. *Woonbeleving Allochtonen*, 2001. www.smo-ov.nl/pdf/onderzoek%20woonbeleving%20allochtonen.pdf (accessed 23 April 2003).

Waterson, Roxana. 'Enduring landscape, changing habitus: The Sa'dan Toraja of Sulawesi, Indonesia'. In *Habitus: A Sense of Place*, edited by Jean Hillier and Emma Rooksby, 334–355. Aldershot: Ashgate, second edition, 2005.

Chapter 4

Memory, Belonging and Resistance: The Struggle Over Place Among the Bedouin-Arabs of the Naqab/Negev[1]

Safa Abu-Rabia

Introduction

The literary critic, Anton Shalhat, notes that the Palestinian dream is the home on the other side of the border: "A person usually lives in a particular place, while among the Palestinians the place lives in the person".[2] This statement explains the powerful connection between the Palestinians and their land, based on kinship with the direct rightful owners of the land.[3] Rubinstein writes that "the Palestinians perceive their lost homeland in the most direct, tangible, and simple of terms: as a field, an olive tree, a balcony, a well".[4] The traditional, localized sense of belonging that was for them the main source of their strength, is embodied by a deep attachment to their homes and villages and reinforced through the documentation of their lives in their destroyed villages and lost lands, not just as it was in the past, but also as it is today, with an emphasis on dispossession and displacement.[5]

---

1   Another version of this chapter has been published in: Ismael Abu-Saad and Oren Yiftachel, eds. *HAGAR Studies in Culture, Polity and Identities, 8/2 Special Issue: Bedouin-Arab Society in the Negev/Naqab: Studies in Policy, Resistance and Development* (Beersheba: Ben Gurion University, 2009), 93–120.

2   From Anton Shalhat, "At the Mouth of the Volcano", *Liqa'a-Mifgash*, 7–8 (1987), 72. Cited in Dani Rubinstein, *The Fig-Tree's Embrace: The Palestinian Right of Return* (Jerusalem: Keter, 1990), 12 [Hebrew].

3   As Morphy writes, land that is full of the memories of the people who lived on it becomes a source of reference for feelings from present-day life. Therefore the land-oriented memory is part of the people's memories. This connects between the people on both a personal and cultural level, while the land is the source: the father's blood ties with the land give the second generation the "permission" to be present on it and the feeling of direct ownership. In Howard Morphy, "Landscape and the Production of Ancestral Past", in *The Anthropology of Landscape*, ed. Eric Hirsch and Michael O'Halnon (Oxford: Clarendon Press, 1995), 200.

4   Rubinstein, *Fig-Tree's Embrace*, 15.

5   Susan Slymovics, *The Object of Memory: Arab and Jew Narrate the Palestinian Village* (Philadelphia: University of Pennsylvania Press, 1998), 1–30.

The struggle over place in Israel/Palestine has two main axes. The first is the construction of a Jewish, Zionist national hegemony over the space that excludes and erases Palestinian history through institutionalized discourses and practices. The second is the Palestinian Bedouin-Arab construction of the connection to their land through spatial practices of pilgrimage, location of old traces, demarcation of borders and recounting the past.

This chapter examines the "struggle over place" as expressed by the Bedouin Arabs of the Naqab/Negev. First I will frame the land conflict in Israel as a manifestation of the struggle over "contested spaces", with all that suggests regarding the planning and organization of the space. Then I will analyse the spatiality of Zionist ideology in the Negev and the Bedouin Arabs' exclusion from it through a description of the Israeli institutional legal and planning mechanisms for "Judaizing"[6] the Negev. Finally, I will describe the counter-process of the construction of a "sense of place" among the Negev Bedouin Arabs and its implications for identity-building as exiles in their homeland.

Research Population

The research population includes 16 Bedouin men of the 1948 generation, and their sons. The informants came from eight families from the Negev,[7] seven of which owned land prior to 1948. Today, five of these families[8] live in recognized villages; another family lives in an unrecognized neighbourhood of a recognized town, away from its historical land; and the other two families live in unrecognized villages.[9] Of these last two families, one belongs to the

---

6   The process of "Judaization" upon which the settlement of the State of Israel is based, is driven by the worldview that sees the Israeli/Palestinian space as "belonging" to the Jewish communities around the world, and the purpose of the state as to concentrate all of these communities in the one territory. The definition of the state as "Jewish" – and not "Israeli" – granted legitimacy to a broad policy of Judaization that anchored the "legal" discrimination against the Arab citizens of the State in legislation and government action. From the definition of "ethnocracy" in Wikipedia, http://he.wikipedia.or/wiki (accessed on 1 September 2009).

7   Research constraints limited the choice of informants mostly to men. However, the effect of these spatial political processes on Bedouin women is currently being investigated in the framework of an ongoing doctoral study. For additional research addressing this issue, see Henriette Dahan-Kalev, Niza Yanai and Niza Berkowitz, eds. *Women of the South, Space, Periphery, Gender* (Tel Aviv: Xargol Press, 2005); Tovi Fenster, *Gender, Planning, and Human Rights* (London: Routledge, 1999); Haim Yacobi, *Constructing a Sense of Place: Architecture and the Zionist Discourse* (Aldershot, UK and Burlington, VT: Ashgate, 2004).

8   The "family" refers to extended and nuclear families that populate the divisions and sub-tribes making up the tribe (Aref Abu-Rabia, "Bedouin Family in the Negev and Livestock Breeding", (Ph.D. dissertation, Tel Aviv University, 1991, 41–43)).

9   The recognized villages were established by the state with the end of the military government in order to concentrate the Bedouin Arabs in them. About one half of the Bedouin

group whose unrecognized villages are on their historical lands; the other family lives far from its historical land and is among those considered internally displaced citizens.

Methodology

The choice of informants was based on spatial and tribal[10] changes that took place after 1948, which were crucial for shaping the new geographic-tribal reality in the Negev space. Most of the Bedouin Arabs who remained in the Negev after 1948 belong to the Tiaha clan, while a few belong to the Al-Trabin and Al-'Alazazma clans. These tribal changes were a deciding factor in the choice of the study's informants as they represent social divisions deriving from tribal-territorial hierarchies between recognized and unrecognized villages, and within them the tribal hierarchies that determined land ownership (or lack of) according to pre-1948 criteria.[11]

Arabs of the Negev live in these recognized villages, while the other half live in the 46 villages in the Negev known as unrecognized villages, which lack in basic infrastructure and services (connection to water and electricity grids, as well as health, education and welfare services). For more on this, see Aurli Almi, "In No Man's Land: Health in the Unrecognized Villages of the Negev" (Tel Aviv: Physicians for Human Rights and the Council of Unrecognized Villages, 2003), 7–10 [Hebrew]; Shlomo Svirski, "Invisible Citizens, the Government Policy Towards the Bedouin in the Negev" (Report for the Adva Center, Tel Aviv, 2005), 24.

10   The term "tribe" or "tribal" has contrasting connotations. On the one hand, researchers define tribe (a'shira) as an administrative unit connected by kinship clans that includes several families (Aref Abu-Rabia, "Bedouin Family in the Negev and Livestock Breeding", 41–43; Sasson Bar-Zvi and Yosef Ben-David, "The Negev Bedouin in the 1930s and 1940s as a Partially Nomadic Society", *Studies in the Geography of the Land of Israel*, 10 (1978), 110–121; Ahmad Abu-Hussa, *Encyclopedia of Bear ElSabea' Clans and its Major Tribes* (Amman: The Middle-Eastern Company Press, 1994), 29–34 [Arabic]). On the other hand, territorial divisions such as tribal-based organizational units were forced on the Bedouin Arabs by colonialist interests strengthening control over them. This control system, which began during the Ottoman regime, was reinforced under the British Mandate and continued during Israeli rule. See Tovi Fenster, "Participation in the Settlement Planning Process: The Case of the Bedouin in the Israeli Negev" (Ph.D. dissertation, The London School of Economics and Political Science, 1991), cited in Morton H. Fried, *The Notion of Tribe* (Menlo Park, CA: Cummings, 1975), 112; see also Leon Shaskolsky Shelef, *A Tribe is a Tribe is a Tribe – on Changing Social Concepts and Emerging Human Rights* (Tel Aviv: Tel Aviv University (mimeo), 1990), 6. In addition, researchers point out that the terms "tribe" and "tribalism" have been replaced in the literature by alternative terms like "indigenous people", "Fourth World people" and "ethnicity" (Ibid.).

11   The transfer of Bedouin-Arabs to the Restricted Area (Ezor Siyag) created a changed geography after 1948. The state placed tribes into new localities among Bedouin Arabs who had resided in the area prior to the establishment of the state, sometimes on land whose ownership was claimed by the latter. Oren Yiftachel, *Land, Planning and the Arabs in the Negev* (Beer Sheva: Ben Gurion University of the Negev, Center for Bedouin Studies, 1999), 9–10 [Hebrew].

The research field focuses mainly on members of the generation of 1948, whose memories inform us firsthand about their lives as it used to be on their ancestral lands, prior to their forcible departure to the space they live in today. Research methods included in-depth interviews with these older Bedouin Arabs as well as family members, especially their children, who offered their points of view. In addition, I conducted participant observation in the diverse ways of Bedouin Arab life, in both recognized and unrecognized villages. These methods allowed me to penetrate the world of the Bedouin Arabs, into the factors that shape their connection to the past, their present life, and the implications of both of these for the future.

The Struggle Over Place

The spatial struggle in Israel is one between "contested spaces".[12] It involves social conflicts in sites of confrontation over control of resources, determination of the place's future in relation to its past, and its cultural meaning over time. According to Kuper, this rivalry between "contested spaces" is expressed in three main areas:

1. *State hegemony and commemoration of sites*. As the conflicts are located in a place where memory is in the process of being constructed, the production and reproduction of hegemonic authority demands a monopoly over public spaces in order to determine the dominant narrative, while weakening contradictory histories of peasants, women, the working class, and others.
2. *Tourist sites*. These sites guide visitors via visually attractive elements that are marketed within the framework of national, economic and political institutions, far from the control of the local inhabitants and in opposition to their local culture. These tourist sites play a critical role in constructing and legitimizing the ideological hegemony of the state.
3. *Identities of places and the politics of representation*. Social identities shaped in terms of a specific place or site construct territorial identities that play an important role in the struggle over place.[13]

---

12 The term "contested spaces" has been researched in the anthropological context as part of the general research on place and space and its socio-cultural associations (Hilda Kuper, "The Language of Sites in the Politics of Space", in *The Anthropology of Space and Place*, ed. Setha M. Low and Denise Lawrence-Zuniga (Malden, MA: Blackwell, 2003), 247–263; Gary Wray McDonogh, "Myth, Space, and Virtue: Bars, Gender, and Change in Barcelona's Barrio Chino", in *The Anthropology of Space and Place*, ed. Setha M. Low and Denise Lawrence-Zuniga (Malden, MA: Blackwell, 2003), 264–283; Steven Gregory, "Black Corona: Race and the Politics of Place in the Urban Community", in *The Anthropology of Space and Place*, ed. Setha M. Low and Denise Lawrence-Zuniga (Malden, MA: Blackwell, 2003), 284–298.

13 Setha M. Low and Lawrence-Zuniga Denise, "Locating Culture", in *The Anthropology of Space and Place*, ed. Setha M. Low and Denise Lawrence-Zuniga (Malden, MA: Blackwell, 2003), 22–24.

Na'ama Meishar defines the Israeli space as "racinated and conflicted".[14] As such, it is constructed along two contradictory axes. The first represents Israeli hegemony, motivated by Zionist logic to control the space exclusively for Jews and excluding Palestinians through official channels. The second, no less intensive, axis is that of the space within the Palestinian consciousness, which rebuilds a sense of belonging based on the identity of the land and its people, through formal and informal spatial practices alike.

According to Fenster, the process of "Judaization" of the Israeli/Palestinian space and erasure of the Palestinian past is achieved through mechanisms established by legal systems of ownership in order to legitimate Jewish-Israeli sovereignty over space.[15] Yiftachel describes this process as being manifested in the Israeli real estate regime, which worked following the establishment of the State to transfer land from a personal, family or village ownership pattern to a Jewish-national one. The two-fold mechanism entailed the confiscation of all property of refugees (including internal refugees) and transfer of ownership of any land classified by the British as *mawat* (uncultivated land) to the State. From 1948 onward, this process took place in the following stages:[16]

1. Unilateral expropriation of land by legal and military authorities, often without legal backing or through emergency laws. The expropriation was made possible through the Absentee Landlords legislation of 1950, which referred to any property owner who left Israel between 29 November 1947 and 1 September 1948 and went to an enemy country; and the Real-Estate Purchase Law of 1953.
2. Declaration of the land as "state land"; limiting land usage via regulations regarding building, agriculture and pasturing; alteration of municipal boundaries; and controlling land allocation for public use.
3. Transfer of land ownership from the British mandatory government to the Israeli State.
4. Transfer of Arab-owned land to the State of Israel, the Development Authority (established in 1952), and the Jewish National Fund. These bodies fall under a single authority, the Israel Lands Administration, which owns 93 per cent of real estate in Israel.
5. Legislation guaranteeing that land ownership will stay forever in the hands of Jewish collective and national institutions, prohibiting any sale of land (Israel Lands Law, 1960).

---

14   Naama Meishar, "Fragile Guardians: Nature Reserves and Forests Facing Arab Villages", in *Constructing a Sense of Place: Architecture and the Zionist Discourse*, ed. Haim Yacobi (Aldershot, UK and Burlington, VT: Ashgate, 2004), 321.

15   Tovi Fenster, "Memory, Belonging and Spatial Planning in Israel", *Research and Theory*, 30 (2007), 189–212.

16   Yiftachel, *Land, Planning and the Arabs in the Negev*, 1–16.

6. Discriminatory sectoral allocation of lands by arrangements between the state and various population sectors: kibbutz, moshav (cooperative Israeli settlement), community settlement, public housing, city neighbourhood and Arab settlement.[17]

In addition, the State uses modern planning tools to enforce the spatial Zionist ideology, by "cleansing" the Israeli landscape of the remains of Palestinian settlement,[18] thus turning planners into implementers of the national agenda.[19] The design of Israeli space is also effected through the commemoration of Jewish sites of memory through numerous state mechanisms, including The Supreme Commission for Preserving Holy Historical Places and Monuments in Israel, founded in 1950; the Society for the Protection of Nature, founded in 1959 following legislation limiting Palestinian pasturing; and the Green Patrol, founded in 1977 to enforce that law. The basic axiom of commemoration in Israel is based on spatial discrimination that defends "lands of the biblical Forefathers" by protecting the ecology of the place from Palestinians "intruders". Thus, the national Jewish landscape is designed by constructing the Arab as the enemy of nature and the landscape, and by the need to preserve nature from Palestinian cultural traces. According to Meishar, this process also becomes a means to mark the borders of Arab villages, to limit and supervise them, as another useful strategy for controlling the land.[20] In addition, the culture of Judaizing space is reinforced in the education system and by youth movements that produce an Israeli landscape pure of any Palestinian presence.

The Bedouin-Arab Context

From the Zionist standpoint, the Negev comprises the largest territory within Israel, the area least populated by Jews, and the last land reserve for implementing its aims and preventing the advent of a "bridge of villages" between Gaza and Hebron.[21] For the Bedouin-Arabs, the land has a social, cultural and economic value that defines their personal and collective identity and status. Their connection to it is sacred, reflecting its acute importance for them in all aspects of their lives.[22]

At the end of the 1948 war, Israeli state authorities began to use legislation in order to transfer Bedouin-Arab land to the state. Anchored in the formal legal

---

17   Ibid.

18   Fenster, "Memory, Belonging and Spatial Planning in Israel", 189–212.

19   Tovi Fenster, "On Belonging and Spatial Planning in Israel", in *Constructing a Sense of Place*, ed. Haim Yacobi (Aldershot: Ashgate, 2004), 285–302.

20   Meishar, "Fragile Guardians", 310–314.

21   Yosef Ben-David, "The Land Dispute Between the Negev Bedouin and the State: Historical, Legal and Current Perspectives", *Karka: The Journal of the Land Use Research Institute*, 40 (1995), 31–38 [Hebrew]; Fenster, "Participation in the Settlement Planning Process", 117–123; Yiftachel, *Land, Planning and the Arabs in the Negev*, 23.

22   Safa Abu-Rabia, "Exiled in our Homeland: The Diaspora Identity of the Bedouin in the Negev" (Masters thesis, Ben Gurion University, 2005), 33–58.

system, such laws allowed government authorities to remove the Bedouin Arabs from their land and undermine their ownership claims.[23]

Military rule, one of the first mechanisms to be employed, entailed the forcible concentration of 11,000 Bedouin Arabs in the Negev within the Restricted Area (*Ezor Ha-Siyag*), north and east of Beer Sheva, which remained under military government until 1966.[24] This mechanism not only regulated their movement and disconnected them from the remaining Palestinian population, but it uprooted them from their land and prevented them from returning to it.

In parallel, the state embraced laws and regulations from the Ottoman legal system, among them the Mawat Law of 1858[25] and the British Mandatory Dead Land Order of 1921. Thus, all Bedouin-Arab land was defined as uncultivated and classified under state ownership.[26] In 1953, the Real Estate Acquisition Law was legislated, expropriating any land that was unsettled and uncultivated as of 1 April 1952. Since most of the Bedouin Arabs were forcibly evacuated from their land before this date, they lost their land rights even though they had ownership certificates.

Another mechanism for undermining the Bedouin-Arabs' connection to their land was an urbanization phase involving the establishment of seven Bedouin townships in the Negev.[27] Today these townships suffer from poverty, unemployment, crime, social tensions and a lack of economic infrastructure. All of these factors turn them into distressed localities, isolated from Jewish villages. In actuality, only about 52 per cent (83,000) of the Bedouin-Arab population moved there, while the remaining 47 per cent (76,000) live in unrecognized villages lacking in economic and social infrastructure. The constant threat of house demolitions in these villages is aimed at making the Bedouin leave their ancestral lands and move to the townships.[28]

---

23   Gazi Falah, "The Spatial Pattern of Bedouin Sedentarization in Israel", *GeoJournal*, 11, no. 4 (1989), 361–368; Fenster, "Participation in the Settlement Planning Process", 125–127; Shlomo Svirski, "Invisible Citizens: the Government Policy Towards the Bedouin in the Negev", Report for the Adva Center, Tel Aviv, 2005, 5; Yiftachel, *Land, Planning and the Arabs in the Negev*, 7–12.

24   Yosef Ben-David, "The Bedouin of the Negev", *Idan*, 6 (1986): 93–99 [Hebrew]; Yiftachel, *Land, Planning and the Arabs in the Negev*, 8.

25   The law defined land as mawat if it was not held or used by people continuously, or had not been allocated by the governor to any local authority and was located away from a known village. These conditions allow the governor to transfer the lands under its authority. From Avinoam Meir, "An Alternative Examination of the Roots of the Land Conflict in the Negev between the Government and the Bedouins", *Karka*, 63 (2007), 22 [Hebrew].

26   Yiftachel, *Land, Planning and the Arabs in the Negev*, 9.

27   Tel-Sheva, Rahat, Kseifa, A'rara, Shgeb-al-Salam, Hura and Laqya.

28   Harvi Lithwick, *Policy Directions for the Revitalization of Bedouin Settlements* (Jerusalem: Center of Social Policy Studies, 2002), 7–26 [Hebrew]; Robi Nathanson, "A Comprehensive Plan for Addressing the Problems of the Negev Bedouin" (Tel Aviv: The Center for the Legal and Economic Studies in the Middle East, 1999), 5–19. See also NGO reports such as: Anonymous, "All's Fair: The Destruction of the Agricultural Crops of

Finally, several bodies have been established to limit the spatial movement of Bedouin Arabs and concentrate them in one area. These include the Israel Lands Administration, the Bedouin Administration, the Green Patrol, the Ministry of the Interior's National Building Supervision Unit, and the special police unit "Rotem", which was established to deal with crime among the Bedouin Arabs in the Negev.[29]

Israeli State Legal and Planning Strategies for Severing the Bedouin-Arabs' Connection to their Land

The land dispute in the Negev represents opposing perceptions, discourses and practices of space between the Jewish and the Bedouin-Arab populations. The state uses its central tools – primarily the legal system and modernist planning discourse – to organize the space of the Negev under Zionist control and Jewish hegemony and to limit the Bedouin-Arabs' hold on their lands. In parallel, academic research and the media contribute to the construction of the image of the Bedouin Arabs as lacking in lands and as a "demographic threat".

Legal discourse in Israel constructs the Zionist enterprise as a moral narrative aimed at "redeeming" the land. In the case of the Negev, this redemption is from the Bedouin Arabs, who are in turn constructed as the immoral, nomadic, chaotic "other", who is invisible in the eyes of the law.[30] This legal discourse is expressed in precise signs: document times and dates that create "facts" for the legislator that serve his "just" claim. In this construction of reality, the Bedouin-Arabs' traditional and unwritten claims of land ownership based on prior conquest are delegitimized.[31]

Simultaneously, there exists a discourse that looks critically at the legal worldview and its construction of the Bedouin Arabs in such a way as to enable the state to dispossess them of their territorial rights. In his innovative research, Avinoam Meir analyses the British Mandatory adoption of, and the subsequent Israeli implementation of, the Ottoman Mawat Law of 1858, showing its irrelevance to Bedouin-Arab culture and contradiction of their traditional ways

---

Bedouin Citizens of the Negev by the State by Chemical Dusting from the Air" (Report of the Arab Association for Human Rights, Nazareth, 2004), 1–5 [Hebrew]; Almi, "In No Man's Land", 4–10, 72–74; "Arab-Bedouin Education in the Negev: In the Shadow of Poverty" (Report of Shatil, the Regional Council of Unrecognized Villages, Beer Sheva, 2003), 2–12 [Hebrew]; Svirski, "Invisible Citizens", 3–31.

29 Yiftachel, *Land, Planning and the Arabs in the Negev*, 10–11.

30 Ibid., 2–16; Ronen Shamir, "Suspended in Space: Bedouins under the Law of Israel", *Law and Society Review*, 30 (1996), 231–241.

31 According to the Bedouin, the unwritten "Hagar Law" establishes their ownership of land they conquered by force. They base the claim of ownership in relation to their ancestors who conquered the land decades earlier and cultivated it continuously since then. In Fenster, "Participation in the Settlement Planning Process", 123.

of spatial organizing space.[32] According to Meir, this law represents a Western-Orientalist perception of "settlement" that applies to sedentary cultures. In this worldview, a settlement is an entity with a defined sprawl in which a collection of people and their activities are integrated, and which is recognized under the formal definitions of a town or rural village.

Another major tool for the dispossession of the Bedouin Arabs from their lands is the modernist planning discourse that designs space in service of national ideology. In the context of the Negev, this tool served to concentrate the Bedouin Arabs into as reduced a space as possible around defined localities and to disconnect them from the traditional lands that define their territorial identity.[33]

Meir offers an alternative Bedouin interpretation of the notion of settlement, in which the ownership of each locality was recognized by the whole Bedouin-Arab public and the areas were inhabited according to the change of pasture season. In these informally organized sites, the inhabitants carried out specific economic and social activities, as well as social relationships full of cultural meaning based on local memory and history. In this context, grassroots attempts at alternative planning options for Bedouin-Arab settlement have arisen. These independent, empowerment planning initiatives[34] are based on principles of spatial justice, diverse types of spatial organization, and cultural pluralism. This planning approach is a social and cultural process that consists of local and indigenous knowledge and incorporates such central aspects as: the identity aspect (the meaning of place in Bedouin-Arab culture as a local-territorial identity), the social aspect (the influence of the Bedouin-Arab way of life on its spatial organization), and the spatial aspect (in relation to Bedouin-Arab village construction, meanings and needs for organizing their lives).

Parallel to the advent of this alternative local Bedouin-Arab planning approach, a no-less-intensive informal local axis has developed, which is aimed at the construction of the consciousness of the place. This next part of the chapter will therefore deal with the ways in which the Negev Bedouin-Arabs construct a "sense of place" on their ancestral lands – through spatial practices that strengthen territorial identity based on notions of belonging and memory. These practices transform their ancestral lands into sites of memory for pilgrims, while their current living spaces have developed into sites of resistance and protest.

---

32 Meir, "An Alternative Examination", 14–34.

33 Fenster, "Memory, Belonging and Spatial Planning in Israel", 189–212; Avinoam Meir, *From Planning Advocacy to Independent Planning: The Negev Bedouin on the Path to Democratization in Planning* (Beer-Sheva: Negev Center for Regional Development & Ben-Gurion University of the Negev, 2003), 5–10.

34 Empowerment planning initiatives, which include such features as objection, durability and re-construction, point to powerful processes among the Bedouin-Arabs in the Negev. On the one hand, this position reflects their needs and socio-cultural structure in the spatial planning, and on the other hand, requires the planning establishment to allow for independent democratic planning. For additional reading about types of planning among the Bedouin Arabs in the Negev, see Meir, *From Planning Advocacy to Independent Planning*, 43–55.

Memory, Belonging and Land: Constructing a Sense of Place

"Sense of place", as defined by Yacobi, is the subjective human emotional attachment that people feel towards a specific place. It is located at the intersection of three main dimensions: a given space, meaning the physical space and the way it is organized; the conceived space, or the way in which planners and architects represent space; and the ideological space, which relates to the evolution of a place as influenced by its socio-political context.[35] Massey adds that the connection between identity and place derives from the need to belong to a specific place that gives one a natural sense of comfort, security and refuge.[36]

According to de Certeau, one gives meaning to a space by walking through it; it is a kind of territorial demarcation based on the acquisition of knowledge, memory and intimate experience in the place.[37] Fenster defines "sense of place" as the connection between past events, such as childhood experiences, and specific places. This connection constitutes the essence of belonging, giving places meaning through social positions and processes of representation.[38]

The conceptualization of the past through the concretization of place has been the subject of a great deal of research with regard to the Palestinians. It has been said that the picture of a Palestinian pointing his finger at the remains of former houses and lands or the village where his parents lived, transports the idea of Palestine from the realm of the abstract to the real by arousing feelings towards the place and a better understanding of the life of the past. An encounter with the physical remains perpetuates the past in the present, while those who visit get the sense that these memories belong to them.[39]

Whereas the Palestinians in the Galilee and the Triangle work to revive the memory of the Nakba so that it will not be forgotten by the younger generation,[40] the Bedouin-Arabs in the Negev continue to live in an impermanent reality and therefore experience the implications of the Nakba on a daily basis. The scar of spatial change refuses to be forgotten and healed and continues to significantly shape their future needs. Thus, the Nakba in the Bedouin-Arab context is not only a memory for the generation of 1948, but the exile-identity of the second and third

---

35   Yacobi, *Constructing a Sense of Place*, 3–7.

36   Doreen Massey, "The Conceptualization of Place", in *A Place in the World?*, eds Doreen Massey and Pat Jess (Oxford: Oxford University Press, 1995), 89–90.

37   Michel de Certeau, *The Practice of Everyday Life* (Berkeley, CA: University of California Press, 1988), 97–99.

38   Fenster, "Memory, Belonging and Spatial Planning in Israel", 189–212.

39   Efrat Ben-Ze'ev, "Transmission and Transformation: The Palestinian Second Generation and the Commemoration of the Homeland", in *Homelands and Diasporas*, ed. Andre Levy and Alex Weingrod (Stanford, CA: Stanford University Press, 2005), 123–137; Maurice Halbwachs, *On Collective Memory* (Chicago: University of Chicago Press, 1992), 46–52; Slymovics, *Object of Memory*, 1–30.

40   Ben-Ze'ev, "Transmission and Transformation", 123–137; and Slymovics, *Object of Memory*, 1–28.

generations – and probably the fourth as well. Concretized in their current identity of impermanence, the Nakba shapes their aspirations to return to their land and is expressed in their everyday lives as exiles in their homeland.

The sense of place of Bedouin Arabs in the Negev is constructed by the 1948 Nakba generation, primarily through visits to their former lands – a symbol of the life of the past. These visits are usually made in the springtime, and on Nakba Day (May 15): "We visit our place, our houses are still there, and our graves are there ... in order to get to know the area where we lived ... our parents show us how they lived. They explain to us how to find the place and show us: we lived here, the well was here, and here we would sit at night ... memories".[41] The importance of developing a sense of emotional ownership is expressed in the visit: "They visit the land ... They feel the land, they kiss it. Oh how they feel and sense the land and become attached to it".[42]

The detailed descriptions and stories about their past way of life on the site are another practice for the creation of the feeling of intimacy that, according to Halbwachs, bonds a person to his land in the framework of the social circle that belongs to that same past.[43] The reconstruction of the past is carried out by locating remnants that, to paraphrase Lowenthal, bear perpetual historical witness to their existence on their land.[44] Says Al-'Azazma, "Every year we go to visit my grandfather's house there, the old houses there. Every year in the springtime we take the children and recount our memories to them, and sit down on our land. We explain that this is *diarna* [our land and our home]. Here we were born and here we lived on this land, here we planted lentils and here wheat, and here we made a *sadeh* [an earthen dam to catch the rain water] with a *jerafa* [backplow] pulled by a camel, and here we worked on the road. That's how I explain it to my children".[45] The demarcation of the borders of the territory also constructs their belonging to the place. Al-'Uqbi: "this is A's land, and that is B's land. *Benhen lel-watan* [we yearn for our land] so that they will know it".[46]

Additionally, the family members make efforts to preserve "proof" of their ownership of their lands, by preserving old land registry documents that they say prove their presence on their lands. They also keep maps and aerial photographs from the British Mandate era. According to Abu-Shareb: "In my papers it says exactly what the northern borders of A and B's land were, as well as C and D from the south ... I have it all organized".[47]

---

41 Abu-Shareb, interviewed by Safa Abu-Rabia, 20 February 2002.

42 Ibid.

43 Halbwachs, *On Collective Memory*, 56.

44 David Lowenthal, *The Past is a Foreign Country* (Cambridge: Cambridge University Press, 1985), 238–241.

45 Al-'Azazma, interviewed by Safa Abu-Rabia, 22 January 2003.

46 Al-'Uqbi, interviewed by Safa Abu-Rabia, 16 February 2002.

47 Abu-Shareb, interviewed by Safa Abu-Rabia, 20 February 2002.

The Goal of Perpetuating Spatial Memory

The need for and preoccupation with these "realms of memory" is connected to the sense of their loss, writes Pierre Nora, who coined the phrase *lieux de mémoire* nearly three decades ago in the French context.[48] Their reconstruction among the Bedouin Arabs in the Negev preserves the past memory and transfers the land to an "eternal realm of memory",[49] manufacturing what Zerubavel calls an "archive of the present".[50]

The creation of these realms of memory reconstructs the history of the Negev Bedouin and gives it prime ideological significance for their children, who receive a sense of direct connection as secondary landowners: "When we go there, we bring its stones, and bring its sand … It is important to me that even the smallest boy know, know that we once had land. When we take him to the land, his mother tells him – this was your land! Once it was ours and we were expelled from it and the place where we live now is not our land, rather this is our land. We are connected to this land".[51] Redefining the past as an ideal for their children's future is perceived as a way to bring the memory of the land alive in their hearts.[52] Memory, which comes into existence and is shaped by the need for a history of the present, turns the timeless continuity of the past into a reality in today's transient era.[53] In Abu-Rqaieq's words, this "will give them continuity to prove their ownership of the land, no more and no less. I take them to show them where I was born, where I would drink water, where I lived my life. I pass on this information so that they will know why we are not getting our due, and why we are still here without the right to our land. I want them to know that we have been denied our right. They shouldn't forget where they came from, they need to know that they were not born in Tel Sheva, we are not from *halbalad* [this land]. *Asalna* [our origin] and *amlakna* [our property], are in a different place and the government is withholding it from us".[54]

The goal of informing the younger generation is to enable them to prove their ownership: "The papers are title deeds … because this is our land that we planted … our land is empty today, and we have proof that it is our land that we own. It is marked. This is my ownership, so one of these days, when perhaps I will die and

---

48   Piere Nora, "Between Memory and History", *Representations*, 26 (1989), 5–8.

49   Ibid.

50   Yael Zerubavel, *Recovered Roots: Collective Memory and the Making of Israeli National Tradition* (Chicago: University of Chicago Press, 1995), xvii.

51   Abu-Shareb, interviewed by Safa Abu-Rabia, 20 February 2002.

52   Mahmoud Issa, "Decoding the Silencing Process in Modern Palestinian Historiography", a paper presented at the conference "Worlds & Visions, Perspectives on the Middle-East Today. Local and National Histories" (Denmark, University of Aarhus, 1997), 3–4.

53   Badea Warner, "Beyond the Boundaries", in *Women and the Politics of Military Confrontation. Palestinian and Israeli Gendered Narratives of Dislocation*, ed. Nahla Abdo and Ronit Lentin (New York: Berghahn Books, 2002), 111–118.

54   Abu-Rqaeiq, interviewed by Safa Abu-Rabia, 16 April 2002.

their mother, too, this is what they will have left. They need to tell their children that there is land, and I have documents and photocopies. I photocopied those documents a few times and arranged them in my folder. I put all my papers away, in order".[55]

Another reason for visiting the old village is to explain the situation of the Bedouin Arabs of the *Naqab* (Negev) today. An understanding of Arab-Bedouin history flows from a statutory ownership of the land, imparted through the construction of a sense of ownership and pride in self and one's status, as expressed through the words of A. Abu Syiam, the son of one of the interviewees: "This is not the first time that he has told us the history, he tells us this all the time. Why do you think I know it by heart? Because he tells it to us at every opportunity in order to explain to us why we have problems today in Laqiya, why we have no land, why we live today on the land of others. We cannot move forward, and so it hurts … The problem of the land, according to the story, and the way they were expelled from the land, and how they lived after that … Today we have a problem; families around us have land, so why do we have no land? Why do we have no land?".[56]

Protest, Resistance and Exclusion: Strangers in the Space

The uniqueness of the Bedouin exile is expressed in their relationship with the State of Israel, which reshaped the borders and took the Bedouin out of their natural setting in order to redefine the space to suit its needs. This impacted the Bedouin-Arabs physical, social, economic and political situation, creating a sense of physical and emotional alienation among them.

Their exile is expressed on two main planes: the physical plane, in that they are removed from the land which is the focal point of the life they yearn for and idealize, and the emotional plane, through the feelings of estrangement and a sense of incompleteness, an absence of rootedness and belonging. This physical and emotional exile is expressed in practice through the construction of memory around the focal point – the native lands from which they were uprooted.

Subsequently, the estrangement the Bedouin-Arabs feel toward their present living situation and place is their main mode of resistance and protest. Their transient and unstable existence, with about half of the Bedouin-Arab population living in unrecognized settlements and the other half living involuntarily and dissatisfied in the recognized settlements, makes it "easier" for them to experience the emotional and physical connection to the past of life on the lost land.

The ongoing feeling of displacement from their homes, land and birthplace is expressed in the Bedouin-Arabs choice to refer to themselves as "expellees" (*msharadeen*). This term is heard over and over in both recognized and unrecognized Bedouin-Arab villages in Israel, regardless of the conditions under which the residents live. From their point of view, the loss is both mental and

---

55   Abu-Shareb, interviewed by Safa Abu-Rabia, 20 February 2002.
56   Abu-Syiam, interviewed by Safa Abu-Rabia, 10 September 2002.

physical, personal and collective, and it has changed their world forever. This loss means a transient life in their present space, and a sharp drop in status from being landowners to landless, as expressed by Al-'Uqbi: "We are considered expellees, landless, we have no land here".[57] This lack is expressed by their estrangement from their present space, even though their current location is just a few kilometres from their former land, as Al-'Uqbi explains: "We are strangers in our homeland".[58]

Their sense of *ghorba* (exile) is multi-faceted. On the one hand they live in geographic proximity to their original lands, while on the other hand they feel a sense of exile due to its loss. They are connected to their original lands through utopian and idyllic memories and the hope for a better life through the return to their land and to "themselves". As such, the Bedouin-Arabs of the Negev live in exile no less than people who were left outside of the borders of the State of Israel. Their physical presence within the borders of the state does not reflect their inclusion in it, since the new borders of the Israeli space have defined them as strangers. As they themselves say: "We are strangers in our homeland, the big homeland. It affects everyone and we can't do much about it ... But there is something that is connected to me personally, to myself. And that which is connected to me personally, to myself, we lost by force, and we want to go back to it".[59]

The Bedouin Arabs construct their exile-consciousness through three main components: the resistance discourse, the construction of an exile-identity, and the vision of return. On resistance discourse, Wedeen writes that the use of particular language and words is a clear indication of resistance against a power, and the style of speech and the words chosen reveal its existence.[60] In the words of some of my interviewees: "The state has raped our lands and has taken our most precious possession ... Today we are a tribe against the state";[61] and: "The Zionists – every dunam of land for the immigrants that will come ... The Bedouin – land and water. It's the height of racism ... Our land could wait 100 to 150 years for the Jews who *might* come ... it's discrimination ... They look at the Arab citizen one way and the Jewish citizen another way ... it's not the same ... The Jew wants to keep their national lands away from the non-Jewish citizen".[62]

The cultivation of the vision of return to their lands is a significant part of the present reality of the Negev Bedouin-Arabs and it is aimed at inculcating the children with the desire to return to the place their parents left. They invest great effort in building a tangible and inviolable connection with the past. In Al-'Uqbi's words: "We have never stopped dreaming or hoping that we will go back to our lands".[63] The return to the land is perceived as a realistic

---

57   Al-'Uqbi, interviewed by Safa Abu-Rabia, 16 March 2002.
58   Ibid.
59   Ibid.
60   Lisa Wedeen, "Acting as if: Symbolic Politics and Social Control in Syria", *Comparative Studies in Society and History*, 40 (1989), 503–509.
61   Abu-Shareb, interviewed by Safa Abu-Rabia, 20 February 2002.
62   Al-'Uqbi, interviewed by Safa Abu-Rabia, 16 March 2002.
63   Ibid.

vision, and their hope of realizing it is integral to the reality of their circumstances under Israeli control. As Al-'Uqbi says: "We must return at any price, we must return to our homes, especially those of us who agreed to be citizens of the State of Israel, we accepted this on the condition that we would live on our land, in our homes".[64] They strengthen their connection toward it under any regime: "If we get our rights in the Israeli era, it will be good. If not, we'll get it later".[65]

Their other primary expression of resistance and protest of their present situation is the cultivating of an exile identity and the conscious reinforcement of the sense of alienation and estrangement they feel toward their current location: "My connection to the land of my father is stronger than my connection to the place where I was born, since I know that it is not mine. And so I don't build as I should, and I don't make more effort to improve my standard of living, I don't even try, since this is temporary ... Even when they [relatives] visit from Jordan, we go to see our land".[66] Their obvious resistance towards their current way of life is expressed in their actions: "Even if they move me to Hura and allow me to build, I will not build there. We don't want to move to Hura, we want to move to our land, we want a village on the land itself".[67] In particular, it is expressed in their refusal to establish their dwelling in any permanent way: "My father would sit ... and we would talk. I would urge him; let's lay down a floor in this room. He would say – no, we don't want to build on other people's land. This isn't our land, and we don't want it. We have no connection to this land, we can't stay here on other people's land. This land is not ours. We didn't buy it. We want our land, which we lived on for hundreds of years, generation upon generation; our fathers, grandfathers, and grandfathers' grandfathers lived on it for hundreds of years".[68]

As far as the Bedouin-Arabs are concerned, the construction of the return mentality entails every effort to prevent the past from being forgotten, in order to perpetuate their efforts to return in the future to the only place they call home. In the meantime, the transplanted generation uses every available means to reconstruct the past by building positive images of life on their own land, in order to create an enduring, constant sense of unbreakable connection to it among their children. Alongside the emotional connection that preserves the Bedouin-Arabs memory are the physical acts of visitation and documentation of the sites of that past.

Conclusion: Between Belonging and Resistance – The Struggle Over Place

Said writes that no one is absolutely free of the struggle over space. This is an interesting and complex struggle because it is not one over soldiers and

---

64 Ibid.
65 Abu-Shareb, interviewed by Safa Abu-Rabia, 20 February 2002.
66 Al-'Azazma, interviewed by Safa Abu-Rabia, 22 January 2003.
67 Al-'Uqbi, interviewed by Safa Abu-Rabia, 16 March 2002.
68 Ibid.

canons, but over ideas, shapes, imagery and imagination. This struggle becomes all the more forceful when dealing with national space that represents a social and geopolitical order concentrating on the homogenization of a population, its collective identity and its territorial borders.[69] Territory is thus above all a geographical expression of power relations in which contradictory claims to the same territory compete to represent it in its cultural meaning and use history to justify its future.[70]

The construction of territorial belonging emphasizes how "sense of place" highlights differences between competing groups, based on social and political structures of power relations in relation to different aspects of identity, such as class, gender and race. The territorial belonging of the dominant group creates differentiation and spatial borders that are another way of constructing the "other" and thus connect between place, power and identity. This may lead to mutual acts of violence, on the one hand, and on the other hand it may strengthen the alternative sense of place amongst the "other", weakened, groups who are threatened by this process.[71]

The spatial conflict in the Israeli/Palestinian context reflects the ways in which conflicting memories and narratives define borders, presence and belonging. On the one hand, the government and its institutions affirm the Jewish-Zionist hegemony in the space, through legal and planning mechanisms that exclude the Palestinian from it. These mechanisms work towards "flattening" the land and erasing the remains of the historical Palestinian presence in order to erase any evidence of another memory and belonging.

Simultaneously, place continues to exist and be preserved among the Palestinians in Israel. This is the case for the Palestinians living in the north of Israel, but all the more so for the Bedouin Arabs in the Negev, whose impermanent and often unrecognized residential status keeps the memory of the Nakba alive on a daily basis. The identity of place and exile, and the vision of return of the Negev Bedouin-Arabs, is constantly and intensively constructed through spatial practices of memory and belonging. The sense of place that attaches the Negev Bedouin-Arab to his past and strengthens his territorial identity as the owner of the place, is constructed through visits to his ancestral lands, demarcation of its borders, touching the land by walking it, locating remains from the past, and recounting the past to the next generation, turning the lands into a site of memory.

Sense of place for the Negev Bedouin-Arabs exists despite institutionalized actions of planning and legislation. At the same time, it nurtures their identity of exile (*ghorba*) and alienation in their current space and constitutes an important part of their discourse of resistance. This exile identity strengthens the vision of

---

69   Edward Said, *Culture and Imperialism* (New York: Vintage Books, 1993), cited in Yacobi, *Constructing a Sense of Place*, 6.

70   Massey, "The Conceptualization of Place", 100; Anthony D. Smith and Colin Williams, "The National Construction of Social Space", *Progress in Human Geography*, 7, no. 4 (1983), 507–508.

71   Massey, "The Conceptualization of Place", 99–105.

return and constructs it as a substantial practical possibility, also among the second, third, and even the fourth generation.

One of the central repercussions of these practices among the Bedouin Arabs of the Negev is the laying of a foundation for a discourse of recognition of their villages by the Israeli establishment. Thus, through their stubborn endurance of unbearable present-day physical conditions and a socio-cultural construction of the space that disconnects them from their ancestral lands, the Bedouin Arabs are working toward the creation of alternative, indigenous planning and legal mechanisms that represent their needs and demands and that include their memory and history in the construction of the Negev space.

References

Abu-Hussa, Ahmad. *Encyclopedia of Bear El-Sabea' Clans and its Major Tribes.* Amman: The Middle-Eastern Company Press, 1994 [Arabic].

Abu-Rabia, Aref. "Bedouin Family in the Negev and Livestock Breeding". Ph.D. dissertation, Tel Aviv University, 1991.

Abu-Rabia, Safa. "Exiled in our Homeland: The Diaspora Identity of the Bedouin in the Negev". Masters thesis, Ben Gurion University, 2005.

Abu-Saad, Ismael and Oren Yiftachel (eds). *HAGAR Studies in Culture, Polity and Identities, 8/2 Special Issue: Bedouin-Arab Society in the Negev/Naqab: Studies in Policy, Resistance and Development.* Ben Gurion University, 2009, 93–120.

Almi, Aurli. "In No Man's Land: Health in the Unrecognized Villages of the Negev". Report of Physicians for Human Rights and the Council of Unrecognized Villages, Tel Aviv, 2003.

Anonymous, "All's Fair: The Destruction of the Agricultural Crops of Bedouin Citizens of the Negev by the State by Chemical Dusting from the Air". Report of the Arab Association for Human Rights, Nazareth, 2004.

Bar-Zvi, Sasson and Ben-David Yosef. "The Negev Bedouin in the 1930s and 1940s as a Partially Nomadic Society". *Studies in the Geography of the Land of Israel*, 10 (1978): 110–121.

Ben-David, Yosef. "The Bedouin of the Negev". *Idan*, 6 (1986): 93–99 [Hebrew].

Ben-David, Yosef. "The Land Dispute between the Negev Bedouin and the State: Historical, Legal and Current Perspectives". *Karka: The Journal of the Land Use Research Institute*, 40 (1995): 31–81.

Ben-Ze'ev, Efrat. "Transmission and Transformation: The Palestinian Second Generation and the Commemoration of the Homeland". In *Homelands and Diasporas*, edited by Andre Levy and Alex Weingrod, 123–137. Stanford, CA: Stanford University Press, 2005.

Dahan-Kalev, Henriette, Niza Yanai and Niza Berkowitz (eds). *Women of the South, Space, Periphery, Gender.* Tel Aviv: Xargol Press, 2005.

De Certeau, Michel. *The Practice of Everyday Life*. Berkeley, CA: University of California Press, 1988.

Falah, Gazi. "The Spatial Pattern of Bedouin Sedentarization in Israel". *Geojournal*, 11, 4 (1989): 361–368.

Fenster, Tovi. "Participation in the Settlement Planning Process: The Case of the Bedouin in the Israeli Negev". Ph.D. dissertation, The London School of Economics and Political Science, 1991.

Fenster, Tovi. *Gender, Planning, and Human Rights*. London: Routledge, 1999.

Fenster, Tovi. "On Belonging and Spatial Planning in Israel". In *Constructing a Sense of Place*, edited by Haim Yacobi, 285–302. Aldershot: Ashgate, 2004.

Fenster, Tovi. "Memory, Belonging and Spatial Planning in Israel". *Research and Theory*, 30 (2007): 189–212.

Fried, Morton. *The Notion of Tribe*. Menlo Park, CA: Cummings, 1975.

Gregory, Steven. "Black Corona: Race and the Politics of Place in an Urban Community". In *The Anthropology of Space and Place*, edited by Setha M. Low and Denise Lawrence-Zuniga, 247–263. Malden, MA: Blackwell, 2003.

Halbwachs, Maurice. *On Collective Memory*. Chicago: University of Chicago Press, 1992.

Issa, Mahmoud. "Decoding the Silencing Process in Modern Palestinian Historiography". Paper presented at the conference "Worlds and Visions, Perspectives on the Middle East Today. Local and National Histories", University of Aarhus, Denmark, 5–6 December 1997.

Kuper, Hilda, "The Language of Sites in the Politics of Space". In *The Anthropology of Space and Place*, edited by Setha M. Low and Denise Lawrence-Zuniga, 247–263. Malden, MA: Blackwell, 2003.

Lithwick, Harvi. *Policy Directions for the Revitalization of Bedouin Settlements*. Jerusalem: Center of Social Policy Studies, 2002 [Hebrew].

Low, Setha M. and Lawrence-Zuniga Denise. "Locating Culture". In *The Anthropology of Space and Place*, edited by Setha M. Low and Denise Lawrence-Zyniga, 1–48. Malden, MA: Blackwell, 2003.

Lowenthal, David. *The Past is a Foreign Country*. Cambridge: Cambridge University Press, 1985.

Massey, Doreen. "The Conceptualization of Place". In *A Place in the World?*, edited by Doreen Massey and Pat Jess, 45–86. Oxford: Oxford University Press, 1995.

McDonogh, Gary Wray. "Myth, Space, and Virtue: Bars, Gender, and Change in Barcelona's Barrio Chino". In *The Anthropology of Space and Place*, edited by Setha M. Low and Denise Lawrence-Zuniga, 247–263. Malden, MA: Blackwell, 2003.

Meir, Avinoam. *From Planning Advocacy to Independent Planning: The Negev Bedouin on the Path to Democratization in Planning*. Beer-Sheva: Negev Center for Regional Development and Ben-Gurion University of the Negev, 2003.

Meir, Avinoam. "An Alternative Examination of the Roots of the Land Conflict in the Negev between the Government and the Bedouins". *Karka*, 63 (2007): 14–51.

Meishar, Naama. "Fragile Guardians: Nature Reserves and Forests Facing Arab Villages". In *Constructing a Sense of Place: Architecture and the Zionist Discourse*, edited by Haim Yacobi, 303–323. Aldershot, UK and Burlington, VT: Ashgate, 2004.

Morphy, Howard. "Landscape and the Production of Ancestral Past". In *The Anthropology of Landscape*, edited by Eric Hirsch and Michael O'Halnon, 184–209. Oxford: Clarendon Press, 1995.

Nathanson, Robi. "A Comprehensive Plan for Addressing the Problems of the Negev Bedouin". Report of the Center for the Legal and Economic Studies in the Middle East, Tel Aviv, 1999.

Nora, Pierre. "Between Memory and History". *Representations*, 26 (1989): 7–24.

Rubinstein, Dani. *The Fig-Tree's Embrace: The Palestinian Right of Return*. Jerusalem: Keter, 1990 [Hebrew].

Shamir, Ronen. "Suspended in Space: Bedouins Under the Law of Israel". *Law & Society Review*, 30 (1996): 231–256.

Shatil, The Regional Council of Unrecognized Villages. "Arab-Bedouin Education in the Negev: In the Shadow of Poverty". Beer-Sheva, 2003.

Shelef, Leon Shaskolsky. *A Tribe is a Tribe is a Tribe – On Changing Social Concepts and Emerging Human Rights*. Tel Aviv: Tel Aviv University (mimeo), 1990.

Slymovics, Susan. *The Object of Memory: Arab and Jew Narrate the Palestinian Village*. Philadelphia: University of Pennsylvania, 1998.

Smith, Anthony D. and Colin Williams. "The National Construction of Social Space". *Progress in Human Geography*, 7, no. 4 (1983): 502–518.

Svirski, Shlomo. "Invisible Citizens, the Government Policy Towards the Bedouin in the Negev". Report for the Adva Center, Tel Aviv, 2005.

Warner, Badea. "Beyond the Boundaries". In *Women and the Politics of Military Confrontation: Palestinian and Israeli Gendered Narratives of Dislocation*, edited by Nahla Abdo and Ronit Lentin, 111–118. New York: Berghahn Books, 2002.

Wedeen, Lisa. "Acting as If: Symbolic Politics and Social Control in Syria". *Comparative Studies in Society and History*, 40 (1998): 503–523.

Yacobi, Haim (ed.). *Constructing a Sense of Place: Architecture and the Zionist Discourse*. Aldershot, UK and Burlington, VT: Ashgate, 2004.

Yiftachel, Oren. *Land, Planning and the Arabs in the Negev*. Beer-Sheva: Ben-Gurion University of the Negev, Center for Bedouin Studies, 1999 [Hebrew].

Zerubavel, Yael. *Recovered Roots: Collective Memory and the Making of Israeli National Tradition*. Chicago: University of Chicago Press, 1995.

Chapter 5

One Place – Different Memories:
The Case of Yaad and Miaar

Tovi Fenster

I said in the meeting that the Southern neighbourhood could be divided into two parts. One part stretches to the existing fence – this is the agricultural area [of the village of Miaar], the other part consists of the ruins of [the houses of] Miaar and we shouldn't touch this part because it was an actual village in the past. I then talked about compromise, about building several houses that would be located at the margins of the village's ruins – in the forest. But we have to reach a compromise ... the majority said what is Miaar? You, the leftists ... and they voted for the plan [to build on the ruins of Miaar]. Two others said that it's impossible to build on the ruins of Miaar. At the end seven voted against the plan and five voted in favour of the plan. (Hana, 28 August 2003)

Negating the reality of agonism does not lead to the harmony and consensus of a fully constituted "we", since agonism, if not antagonism, is constitutive of social identity itself. We then tend to misrecognize the true cause of our failure. Thinking that they are missing some kind of "golden rules", planning theorists attempt to find and to follow normative "golden rules" closely. (Hillier, 2003: 51)

Hana's description of the negotiations that took place at the meeting of the Yaad members, about the plan to enlarge their settlement, reflects more than anything else one of the most complicated issues in Israel today – that of Jewish planning and development in areas and spaces which still contain the remnants of Palestinian pre-1948 villages. Such conflicts are not exclusive to Israel, in fact they characterize daily life in many multicultural spaces in the age of globalization. This is the reason why planning theorists and practitioners felt that they needed to develop new approaches regarding the planning and development of multicultural spaces. These planning approaches suggest ways and methods of dealing with such complicated situations and Hillier, in the quote above, relates to one such effort. In this chapter I will be focusing on two approaches – Therapeutic Planning, which was developed by Sandercock (2003) and suggests a method of involving communities in conflict in planning situations, that leads to the appearance of a consensus; and the agonistic approach, developed by Hillier (2003) which uses the antagonistic view of social relations as more relevant to areas of dispute. Hillier

(2003) argues, as quoted above, that consensus and agreement are not realistic or preferred targets of planning processes. Rather, she introduces the view of conflict and disagreement as part and parcel of the planning process.

Planning processes in places of different memories and symbolism for Jews and Palestinians are common practice in Israel. However, the country's official ideological narrative has yet to acknowledge this historical and tragic situation, which is that the construction of the Jewish State in 1948 meant the destruction of the Palestinian settlement structure, which the latter term as the Nakba (disaster). From the beginning of its establishment and as an act of fear and denial of this historical situation, the State of Israel did its best to erase the ruins of Palestinian villages, thinking that this would be a step towards forgetting (Fenster, 2004a, 2007). This situation has changed in the last decade, with the emergence of a new discourse "from below", such as the one reflected by Hana and other residents of Yaad, who called for a public discussion on these matters, and for reaching a compromise between past and present. This chapter focuses on the dynamics of remembering and forgetting, memory and belonging, at the local level of planning. My aim is to deal with the question of whether it is possible to commemorate the Palestinian past, while developing the Jewish present, in a situation of consensus between these two communities or whether this situation is unsolvable and therefore it is necessary to learn how to live with constant conflicts and disagreements and see the positive impact of conflict, as suggested by the agonistic approach. I will be looking at the social and political processes which took place in Yaad-Miaar, while referring to the wider political and planning context in Israel.

Historically, the design of spaces in Israel since its establishment involved two contrasting processes: the development and building of the Jewish spaces and the destruction and erasing of the Palestinian spaces, as part of formal and informal policies. These policies are part of a wider apparatus of discrimination and oppression of the Palestinians citizens of Israel.[1] It is important to emphasize that planning serves as an efficient tool in order to realize this wider programme, which is based on the Zionist ideology of Judaizing the Israeli landscape. This includes clearing the landscapes of Palestinian "evidence" and practicing ethno-national separation, especially in urban areas. These principles are not unique to the Israeli apparatus, they are based on modernist principles of erasing the old or the past and building "new" spaces, with the help of new technologies and efficient planning methods (Sandercock, 1998, 2003). This desire of modernist planners to construct new and modern spaces in Israel has served the principles of Zionist ideology in three ways: first of all, by rejecting the bourgeois lifestyle; second, by rejecting the lifestyle of

---

1   Much research has been devoted to the oppression and discrimination of the Palestinians in Israel, as well as to issues of identity, memory and contrasting sense of belonging between Palestinians and Jews, in Israel. To name only a few: Anderson, 1999; Fenster, 2004; Golan, 2002; Khalidi, 1992; Moris, 1997; Ohana and Westreich, 1996; Padan, 2004; Saadi, 2001; Shenhav, 2003; Slymovich, 1998; Yiftachel, 2001; Yona and Saporta, 2000; Zerubavel, 1995.

the diaspora; and third, by rejecting the oriental lifestyle. Instead, it has enabled the building of a new socialist and modern landscape (Nitzan-Shiftan, 2000).

These spatial processes were able to take place with the help of three social and political mechanisms developed in Israel since 1948 (Kimmerling, 1977, 1983): presence, ownership and sovereignty. First of all, the physical presence is created. This physical presence then bases itself on the creation of legal apparatuses of ownership, and this legal ownership enhances the sovereignty over the land (Fenster, 2007). Clearing the Israeli landscape of Palestinian ruins can be seen as part of these mechanisms. This project of clearing the landscape of the remnants of abandoned Palestinian villages started in the mid 1960s, with the support of various Israeli governments (Ishai, 2002). One can define these actions as a purification of space, a concept developed by Sibley (1995) and used by Falah (1996) to indicate activities that he terms "evidence clearing". Yacobi (2004) and Yacobi and Zfadia (2004) also mention the use of planning as a means to create "Jewish spaces". One of the expressions of such planning tools are the spatial patterns of segregation between Jewish and Arab citizens in what are termed "mixed towns" – Haifa, Jaffa, Lod and Ramla (Rabinowitz, 1997). Another planning and policy means towards Judaizing Israeli spaces is the project of public housing in Israel. As Kalush and Law Yone (2000) argue, the mechanism of planning and the implementation of public housing in Israel is yet another example of presence, ownership and sovereignty in Israel. The design and construction of public housing in Israel was not merely another housing project for Jewish immigrants following the Second World War, but also a means to enhance Jewish national hegemony and its spatial expressions. Yiftachel (1998, 2000) points to similar processes, when indicating the links between mechanisms of social control of the Palestinian citizens of Israel and urban planning in Israel, which he terms "ethnocracy". These actions clearly show how planning instruments became means for promoting Zionist ideology and for enhancing the Jewish presence, while erasing the Palestinian ruins.

This is also the case in the Galilee, in northern Israel, an area that has been considered as complicated and problematic from the point of view of Jewish–Arab relations, mainly for the following reasons: first of all, part of this region was defined in the 1947 Partition Plan as part of the Palestinian State, and it was only after its occupation in the 1948 war, that the State of Israel declared it a sovereign part of the country. Second, the Galilee was the only area in Israel that had a majority of Arab Palestinian population, despite the fact that 124 of its 190 villages were destroyed after 1948. This situation caused much concern within the Jewish establishment and led, in the 1980s, to the development of a plan to Judaize the Galilee. This plan included the establishment of Jewish settlements, in order to keep the demographic balance between Jews and Palestinians, and to keep the spatial control over the area. The Israeli Planning Authorities consolidated a new type of settlement – the *mitzpe* (Hebrew for "observation"). This refers to its location in high topographies, above Arab towns and villages, with the intention of creating a situation of control. At the beginning, each *mitzpe* consisted of 10–20 families. However, over the years most of them grew into large communal settlements, with some 100–200 families each. All in all, some 66 such settlements were built in the Galilee, each of them covering tens of dunams (Efrat, 1984).

These massive Jewish constructions in the Galilee implied the confiscation of Arab land, which led to massive Arab demonstrations, starting in early 1976. During these events several protesters were killed by the Israeli police and, to this day, these deaths symbolize the bitter relations between Arabs and Jews and are commemorated every year in what is called the Land Day. Another wave of demonstrations, which ended with the killing of 13 Palestinian citizens, took place in the Galilee in October 2000 on the same grounds. These events emphasize the ongoing tension in this region between Jews and Arabs, each side expressing its horror and fear of the other side.

Within this national political context, I will be discussing the local planning developments that took place in Yaad, a Jewish settlement established in 1974 as part of the project of Judaization of the Galilee, and which intended in 2003 to add another 250 houses. This expansion was planned on the land of Miaar, the Palestinian village that existed in the area until 1948. I will first outline the theoretical framework on memory, belonging and planning.

Memory, Belonging and Planning

The growing interest in recent years, in the links between memory, belonging and commemoration, can be seen not only in the political and historical contexts, but also in the social, civil and cultural contexts (Crang, 1998; Mitchell, 2000; Shenhav, 2003; Yuval Davis, 2003). As Pierre Nora (1993) notes, this massive occupation with memory indicates the changes that are taking place in the interpretation meanings of memory. "Lieux de mémoire" or "sites of memory" are created because there are no longer "milieux de mémoire" or "spaces of memory". As he claims, trends such as globalization, democratization and the development of telecommunication put an end to the ideologies of memory, which ensured a smooth passage from the past to the future. It is precisely because we do not live our memories, that we need to allocate sites of memory. These processes of spatialization of memory and the need for tangible signs and representations of memory mean that the real memory is not inside us but within history. These sites of memory – memorials, archives, memorial days and ceremonies – are the means which are created once our spontaneous memory does not exist anymore. In fact, the spatialization of memory makes it part of history (Padan, 2004). The distinction between collective memory and history was made by Halbwachs in the 1950s, when he spoke of the notion of memory and history as separate representations of the past. While he defined history as a scientific learning of the past, which is disconnected from the socio-political reality of the present, he considered collective memory as an organic part of social life, that changes in reaction to the changing needs of society (Halbwachs, 1980). The two notions of memory and history, however, are seen as products of power relations, a fact that many researchers tend to ignore (hooks, 1990).

The research on these notions, as they are found in the field of urban planning, is associated with the power of planning to design urban spaces and to dictate which memory is commemorated, and which is destroyed (Healey, 1997; Hillier, 1998; Jacobs, 1996; Sandercock, 1998). Another area of research is that which acknowledges the value of belonging and its daily deconstruction, thereby challenging in a sense religious types of belonging which are based on the sacralization of sites and places (de Certeau, 1984; Fenster, 2004a; Leach, 2002). In Israel, the formal planning discourse does not relate to these issues yet. Unlike in Australia, the United States, South Africa and other settlers' societies, in Israel there is no official recognition of the different voices which deal with past memories, such as those of the Palestinians. Nevertheless, besides the official ignorance of these issues, some local processes are challenging the formal planning processes, by introducing local planning initiatives. These local initiatives are based on the acknowledgement of the right to commemorate memory and belonging in planning processes, as a component of the right of inhabitants to take part in designing their living environments. Such right is gained by living in the city, as noted by Lefebvre (Lefebvre, 1991a, 1991b).

Four main reasons account for this change in the local planning discourse in Israel, as opposed to conformity in the planning establishment: first of all, the development of international academic knowledge on an alternative conceptualization of planning emphasizing resident involvement in acts of planning, and which focuses on the ways professional and local knowledge can be integrated (Fenster, 2002, 2004b; Sandercock, 1998, 2003). Second, the social and political changes that Palestinian society is undergoing, from being the "surviving generation" to being "a proud generation". These changes reflect a different self-definition of what it means to be an Israeli citizen, and made their struggle for their citizen rights more explicit (Rabinowtiz and Abu Baker, 2002). The third reason for the change in the local discourse in Israel relates to the growing number of NGOs that work to promote issues of citizen and human rights, as part of the development of a civil society in Israel (Ben Eliezer, 1999; Gidron, Katz and Bar, 2000; Ishai 1998, 2002). This includes the establishment of NGOs that deal directly with the promotion of human rights in the field of planning. The fourth reason for the change in the local discourse in Israel is the engagement of the field of human rights with the field of planning. The term "planning rights" is a new definition of this engagement, which legitimizes the use of "equality" and "justice" in the field of planning, rather than in legal studies alone.

Such processes enable the expansion of the planning knowledge beyond its physical vocabulary to include terms such as "memory" and "belonging", or "the right to commemorate memory" and "the right to respect a sense of belonging", as a part and parcel of a whole body of knowledge that is being developed lately, under the title of "planning rights". The right to memory and the right to belong, just like the right to the city, are normative rights that are not given by the establishment but are claimed by the city's inhabitants. The awareness of these rights develops through daily routine life experiences. This change in the planning discourse is expressed in the struggle of citizens, NGOs and civil organizations to incorporate various types of knowledge in the design of Israeli landscapes, all of which relate to notions of memory and belonging, such as gender knowledge, ethnic knowledge, etc. (for more detail, see Fenster, 2002). This change in the planning discourse in Israel reveals the voices from below and allows them to be heard and to become visible; it also helps to incorporate new planning approaches, especially therapeutic and agonistic planning.

The Therapeutic and the Agonistic Planning Approaches

Obviously, the sense of individual and national belonging and the need to commemorate communal memory are constructed independently from the power of planning to maintain or destroy sites of memory. However, one must remember that planning processes do have the power to approve the presence and visibility of mythical sites or to destroy or ignore them (Jacobs, 1996). This connection between remembering, forgetting, city builders and urban planning is being critically investigated in the current planning literature (see for example: Forester, 1999; Healey, 1997; Sandercock, 1998, 2003). Juxtaposed to the dominant role played by modernist planners in commemorating or erasing memory, are the "post-modern" paradigms that do consider notions of memory and belonging as being a part and parcel of the planning process. Hillier (1998), for example, presents such elements in the planning discourse of the Swan Valley in Perth, Australia. This area had a large brewery and has been claimed by the Aboriginals as a sacred site. In the 1980s, several entrepreneurs submitted development plans to build a commercial centre in the area. Hillier presents the different discourses of the city builders involved in the process – the capitalist discourse of the entrepreneurs – and juxtaposed to it the Aboriginal discourse on memory and belonging, as they wished to get the site back and "clean" it from its development by white men. This and many other examples indicate the paradigmatic changes that the planning profession has been undergoing in recent decades. On the one hand, the modernist discourse is still very dominant, emphasizing the significant role of professional knowledge in the planning discourse (Fenster, 2002, 2007). On the other hand, new and different approaches are emerging, that view the planning process as a socially- and culturally-oriented process which perceives the "other", i.e. the indigenous, as a significant party in the planning process, and which considers their local knowledge and sense of belonging and memory as valid. This new vision calls for considering multiple layers of knowledge, rather than a single professional knowledge (Sandercock, 1998). One of the methods that developed from this line of thinking is Therapeutic Planning, coined by Sandercock (2003). She uses this term to indicate the breakthrough in the thinking of how to deal with common spaces in neighbourhoods, cities and regions, where communities of distinctive and different needs live together in a way that can lead to social, value and institutional changes. Such changes are reflected in the openness and understanding that communities in conflict show towards each other. Sandercock bases her analysis on a planning event that took place in a mixed neighbourhood in Sydney, Australia. The city council allocated an area in a mostly white neighbourhood as a residential area for the Aboriginals; in addition, the council gave the Aboriginals the right to manage their residential area. However, due to the presence of drug-related activities among the Aboriginals, and the desire of the city council to "clean" the city from undesirable events prior to the Sydney 2000 Olympic Games, the council decided not to renew the housing contracts with the Aboriginals, to destroy their public buildings and to build instead an open park and a police station. Most of the white residents in the neighbourhood favoured this programme. This plan led to tensions and protests, and divided the neighbourhood residents into three groups: white residents who favoured the plan, white residents who opposed the plan and Aboriginals who opposed the plan. In light of these tensions, the city council decided to recruit a social planner, who initiated a therapeutic process which included three steps, each of which took three months: the first step included meetings between the planner and each of the groups separately on an individual or communal base. The second step included the "speak out" meetings between the groups, in order to establish a safe space for expression and encounter between the groups' members. Only in the third and last step did the planner discuss with the three communities options for the development of the neighbourhood and its public buildings in a way that met the needs of white and Aboriginal residents. This step was based on negotiations between the groups as to the relevant and desired development programmes in the neighbourhood, with the aim of reaching a consensus solution regarding the future of the neighbourhood. The planner managed to identify similar visions of development, and the white neighbours withdrew their objections to a development plan which included the opening of an Aboriginal community centre. It should be noted that the social planner's fee exceeded AUS$50,000.

This case study helped Sandercock identify the Therapeutic Planning process that can be initiated by a planner who is sensitive to the special needs of the communities involved in the process, a planner with the ability to listen, to negotiate and to analyse the situation. The planner is, in fact, a facilitator with a high level of awareness of the needs of individuals and communities. What characterizes this approach is its moderate steps and especially the "speak out" process at the second stage. What makes this planning process a therapeutic one is the ability to create safe spaces and to enable people in situations of conflict to express their fears, desires and hopes without feeling threatened if they speak out. This kind of planning process perceives the past and the communities' history as an important part of the development of future programmes. This approach, argues Sandercock, has the potential for change and for the empowerment of communities. Indeed, the Therapeutic Approach calls for an in-depth perception of the stories, history and emotions of individuals and communities, besides their needs and aspirations in relation to their places. As such, this approach treats equally the history and memories of all parties involved in the planning process, or the white and the Aboriginals, as in the example above. The "speak out" practice at the second stage is what enabled reaching a consensus at the third stage. However, this goal of getting the parties to reach a consensus is, in fact, the basis for the criticism raised against Therapeutic Planning.[2] Hillier (2003) uses the notion of "agonism" to explain her criticism against the idea that a consensus can be reached in such planning processes. "Agonism" is a term also used by Foucault (Foucault, 1982) to define a permanent situation of instability, conflict and constant provocations, a situation in which power relations are not constant or stable but are expressed by an ongoing conflict. For Foucault, criticism is an act of agonism and for Hillier, an agonistic situation within a constant and irreversible conflict is the reason why strategies of consensus – those that Habermas and Sandercock aim to reach – cannot take place in the planning practice. What Hillier argues is that decision-making processes in planning practices must take into account the issues of identity and conflict, as part of the pluralistic character of democratic societies; therefore, the term "agonistic spaces" is more suitable to describe the social change involved in a planning process, than "spaces of consensus", as suggested by the Therapeutic Approach. Agonistic spaces are political and relate to the legitimate and public challenge of the communities' access to resources (Connolly, 1998). Agonistic spaces are characterized by a competition for recognition, priorities and excellence, that create social antagonism (Mouffee, 2002). Hillier (2003) argues that since it is impossible to abolish antagonism there is a need to transform antagonism into agonism, i.e. a situation whereby conflicts can be instructive rather than destructive, and whereby decision-making concerning planning is not necessarily consensual. Therefore, agonism is not only a fight against but also a fight for. That is to say, an agonistic space does not diminish power in the search for a consensus, but accepts power and channels it towards a deeper understanding of people's needs, stories and aspirations. In the next section, I will analyse the dynamic processes that took place between the residents of Yaad and Miaar, using the principles of Therapeutic Planning, and examining the criticism of this approach. However, before doing this, I will introduce the planning history of Yaad and Miaar, within the context of political geographies in the Galilee.

One Place, Different Memories: Yaad and Miaar

The year 1996 marked the government's initiative to Judaize the Galilee by building the *mitzpim* – the Jewish settlements. The communal village of Yaad was established in 1974 as part of this plan. Like most settlements built after 1948 in the Galilee, Yaad too was built on the agricultural land of a pre-1948 Palestinian village: until 1948, some 1,000 Palestinians lived in the village of Miaar, where they owned approximately 10,000 dunams of agricultural land (Zionist Archive, 7/1/43). In July 1948, the village was occupied by the IDF and, according to IDF reports, was emptied of its residents, who fled (IDF Archive, file 100 49/716).

---

2   Similar criticism was raised towards the Habermasian Communicative Approach, which also calls for reaching a consensus in the planning process.

According to the IDF report, the village houses were destroyed during the occupation, although some of the village elders claim that 10 families returned to their homes and lived there until the beginning of the 1950s, and only then were the houses destroyed. Some of the residents escaped to Lebanon, while others moved to nearby Arab villages. Since then, most of them maintain a ritual tradition of conducting a pilgrimage to the graveyards in Miaar, to honour the elders who are buried there. This trend has been enhanced recently with the participation of the young generations (Ibrahim, interviewed on 28 August 2003). These rituals, in fact, represent the connection between the Palestinian past and present and help enhance the sense of belonging among the different Palestinian generations.

de Certeau's (1984) definition of belonging as a sense created out of daily repetitive practices, is very relevant to these practices. It speaks of the way in which picnics and pilgrimages to Palestinians' sites of memory, based on past memories, reconstruct their sense of belonging into a renewed sense of belonging which is based on present, shared and communal experiences, and which turn the past, imagined memory of deportation into a daily, present memory. These daily practices keep the sense of belonging and even enhance the claim for ownership of the place even if it is more symbolic then tangible. Bell (1999) terms these daily repetitive practices as part of performativity – a process which enhances and spatializes a sense of belonging. These repetitive, ritualized practices serve as means for these communities to symbolically appropriate themselves of the territories. Performativity is a reflective act between the individual, the family, the community and the specific symbolic site. Such repetitive rituals construct and enhance a sense of belonging and memory of the past (Leach, 2002). This is how the Palestinians transform the 1948 bitter memory of deportation into a daily, present memory of second and third generations of Palestinians.

When it was founded, the village of Yaad consisted of 100 houses, that were built on the agricultural land of Miaar, and not on the ruins of the houses. This is an important point emphasized by the residents of Yaad – they made sure that the village would not be built on the ruins of the houses, but on its agricultural land. In 2003, the village council expressed its desire to expand it with an additional 250 houses. The village committee suggested the southern alternative, which meant building the 250 houses on the ruins of Miaar. This alternative was approved by the village assembly. A description of this meeting appears at the start of this chapter, in Hana's words, and it is clear that the meeting was not easy. After its approval at the village assembly, the plan was submitted and approved by the local planning committee and was submitted to the Regional Planning Committee, where one can submit objections to a plan. Several bodies and individuals submitted objections to the plan. Among others, the Jewish National Fund and an NGO called Zochrot ("Remembrance"). But the interesting and challenging objections were those submitted by Hana and by the residents of the Palestinian village of Miaar. I would like to focus on these two objections because they represent antagonistic situations that have the potential of becoming agonistic, using the Therapeutic

Planning method. I will first describe the events, and then analyse the painful relations between the residents of Yaad and Miaar.

The first encounters between the Jews and the Palestinians started on the basis of the submission of objections, in the years 2003–2004 (Fenster, 2007). The mutual goal of objecting to the plan to build houses on the ruins of the village created a common ground for personal meetings between some of the Yaad and Miaar residents. In Yaad, the objection to the plan was led by Hana. As she noted, she did not like the plan from the start, because of environmental and landscape issues – the project necessitated the destruction of the forest and of the archeological ruins identified as the remnants of a Jewish settlement from the Roman and Byzantine periods. Hana also objected to the plan for another reason: the project was to be located near her home and would destroy the landscape around it. But, as she emphasizes, the more she got involved, the more she realized that this project could have negative implications, not only for herself but for the entire region. At the beginning she tried to persuade the village assembly, but did not succeed. What made her publicize her objection was her growing awareness that this project could potentially undermine the delicate web of relations between Yaad and the neighbouring Arab villages. As she noted:

> We organized a conference for the NGOs working in the Galilee and I met there my Arab neighbours and I realized for the first time what was going on. In the last year I understood that we needed to act against building our extension on the ruins of Miaar as it could light a big fire ... we have common Jewish and Arabs activities in which we commemorate their Nakba and our memorial day. (Interview, 27 August 2003)

This understanding that the symbolic "Palestinian home" should be respected is the result of meetings between Jews and Arabs that began only after the objections to the plan were prepared. In fact, until the objections were submitted, Hana had not met with the descendents of Miaar, despite the fact that some of them lived in nearby Arab villages. This is because the two communities – Jews and Arabs – that live in such geographical proximity, neither mix nor handle any kind of relations in their everyday life. The objection to the plan was prepared by a small group of Yaad residents, together with Hana, and they were joined by the NGO Zochrot ("Remembrance") NGO. This was a very difficult step emotionally for Hana and the other residents of Yaad, as they were a minority and they were blamed by the majority of Yaad residents for exposing "inner matters" and betraying them. A lot of pressure was put on Hana but, despite all of this, they submitted their objection. The objections on the part of Yaad residents, together with those of Miaar's descendents and of the Zochrot NGO, all deal with issues of memory and belonging. In her objection, Hana suggested that the sites should be preserved and developed:

> The hill [where the new neighbourhood is meant to be built] has been the residential area of the village of Miaar. The decision not to develop this area

was accepted when Yaad was established by the pioneers because of the respect they felt towards the memory of Miaar. Some of the residents are still living with us in the nearby villages. Jewish–Arab relations are very delicate and painful, so that the act of building on the village residential area would be an act of ignorance and cruelty, an ignorance of the disaster that happened to the Palestinians and of their need to commemorate their past. This act (of building a residential site) will harm our efforts to rehabilitate our relations with our Arab neighbours. (Objection to the Yaad plan, 31 July 2003)

The objection submitted by one of Miaar's descendents is also based on the issue of memory and the sense of belonging to the place:

Since the residents of Miaar who live now in nearby villages still visit the site of the village and they do so because of their sense of belonging and attachment to the site and the graveyard, and me and my parents go to visit our relatives who are buried there, it would be very destructive for us if this project is carried out near the ruins of our village and the graveyard. (Objection to the plan, 27 March 2003)

As we can see, Jews and Arabs used a different discourse from the professional planning discourse in their objections. Their discourse is based on expressions that are part of their daily experiences, feelings and emotions of living in the Galilee: "Arab–Jewish relations", "feelings", "a very painful topic", "very delicate relations", "conciliation", "co-existence", "belonging"; all of these expressions relate to daily practices that cannot always can be part of the professional planning discourse which focuses on land size, procedures, location, measurements, etc. These contrasting differences between the two discourses are expressed in the two following quotes. The first one is by the village descendent who talked to the planning committee and explained his objection, and the second one is by the local planning committee engineer:

It hurts me that they built on my home, I don't know what kind of democracy is this which allows the destruction of one home for the building of another one. The house hasn't been destroyed, they destroyed it … if they built for both the Jews and the Palestinians I wouldn't object to the plan.

Because of the many constraints that exist around Yaad it is difficult to reach the targets that the State indicated and there is a need for efficient use of every plot of land that is found within the settlement.

Clearly, each side uses the wording and discourse which expresses their own opinions and interests and the question is how these two types of discourses can be linked. In this particular case, the regional committee accepted the objections partially and only approved the building of some houses, and not of all of them. This result did not satisfy Hana and the other opponents, and they submitted their objection to the

National Planning Committee. In the meantime, more and more residents of Yaad began to doubt the necessity of the original plan and grew to see its problematic nature. This led to a new decision, this time by the Yaad General Assembly, only to build 14 houses on the agricultural area of Miaar. The decision satisfied the descendents of Miaar despite the formal decision of the Regional Planning Committee to approve the building of a larger number of houses in Yaad. In fact the issue was sorted out informally between the two communities. This in itself is a great achievement for both Yaad and Miaar, and in the rest of the chapter I will be focusing on the analysis of the relationship between the residents of Yaad and Miaar in the period following this decision, in order to understand whether what followed can be identified as a Therapeutic Planning process and to what extent there was a need for an agonistic view of relations, rather than for a consensual one.

"I Own the Place. I Have a House, Some Land and I was Uprooted from It" (Nabil, Miaar, 15 October 2004)

This quotation is part of a conversation that took place at one of the meetings between the residents of Yaad and Miaar. The meetings began at the same time as the submission of the objections to the plan, but in fact they became more frequent after the settling of the compromise agreement between them, as detailed above. Hana and Abed (one of the young Arab descendents of Miaar) thought that it was necessary to initiate a series of workshops between the residents of the two villages. This idea gained a lot of support, especially from academics and professionals who specialize in conflict resolution, who raised funds for the workshops, thereby enabling a total of eight meetings to take place between 2004 and 2007 (for details see Babbit et al., 2006). When reading the protocols of the meetings, it is obvious that although the explicit goal was not to discuss planning issues, they were the focus of the discussions. It is therefore interesting to investigate the dynamics of these meetings, from the viewpoint of principles of Therapeutic Planning – antagonism and agonism. I base this work on the protocols of the meetings and on personal interviews with some of the participants.[3]

The meetings took place in Yaad and in Tamra – a nearby Arab village. They included storytelling of the Jewish and Palestinian history, as well as joint walks to the Miaar Hill – where the remains of the village are still found. The walks to the site enabled the Palestinian descendents to tell their stories and to share their common memories but also to relate them to the present practices of pilgrimages and picnic events. On the third walk, the two groups discussed the Yaad extension plan to build houses on the site of Miaar, with particular reference to both the large-scale plan (250 houses) and the compromise plan (20 houses). In a sharing session that followed, one of the Jewish participants talked about the deep emotional experience

---

3   I would like to thank Chassia Chomsky-Porat for sharing with me her experiences and the protocols of the meetings.

that she had when she heard the stories of the Palestinians, especially when they talked about the ruins of Miaar as their home. She noted how important it was for each group to take the other group to their own homes and to discuss feelings, emotions and reflections on past and present memories. Other Jewish participants said that, for the first time, they understood the meaning of deportation and exile, from the symbolic as well as from the physical home. The Palestinians were excited about the possibilities of sharing their pain and suffering but at the same time they expressed their frustration and anger: "You host us but I'm not your guest ... I can't forgive ... it's true that you plan houses there but I also have a key"[4] (Nabil, Miaar, 26 November 2004). Another Palestinian participant indicated that this experience was quite unique for him as it enabled him to visit the people that come "from the nation that caused the disaster we're in now ... I feel anger at the Jewish nation but I don't feel anger at the people that take part in these workshops, but I do ask myself whether they felt [the same pain] that I felt today. I felt a pain when some of them touched the tombs in the graveyard. Some of the new houses touch the ruins of my village. I feel that every stone there is an evidence of our disaster. Every touch at these stones hurts me" (Wafik, Miaar, 26 November 2004).

The meetings' framework obviously did not include three clear steps, as the Therapeutic Planning suggests. The meetings did allow discussions on painful issues from both sides but they did not create structured dynamics similar to those that took place in Sydney between the white and Aboriginal residents, because the purpose of the two types of meetings was different: the one in Sydney was initiated by the municipality and facilitated by a social planner, and the one in Yaad-Miaar was initiated locally, facilitated by conflict resolution facilitators, and funded by external sources (Babbitt et al., 2006). The meetings in Yaad–Miaar did include the sharing of suffering and painful past, but they did not generate a sense of equality and legitimacy with regards to all aspects of history, perhaps because there was no "speak out" step that usually precedes the ensuing agreement.

The meetings showed that Jews hold more racist views of the Palestinians than vice versa. The meetings also clarified the extent to which the Palestinians still emphasize their collective memories and dreams, while the Jews talk about their individual aspirations (Babbitt et al., 2006). At the same time, the meetings exposed the similar feelings of refuge that both Jewish and Palestinian experience either with relations to the Holocaust or the Palestinian disaster – the Nakba. As one of the Jewish participants noted: "The first step has been taken already. To listen to the pain of the other, to empathize and to look for similarities in my own story (of deportation) ... I understand now that the Palestinians have their own painful stories of deportation like I have my painful story of deportation. Like them, I still suffer from my own deportation" (Chassia, Yaad, 31 December 2004).

---

4   The concept of the "key" is a symbol for the Palestinians of their home ownership and of their right to return to the home that they left or they were deported from, in 1948. Most of them left in a hurry with the belief that they would return shortly after. They keep the keys of their 1948 homes to this day and use them in demonstrations and in their public activities.

Indeed, other participants related to this common suffering, from the Holocaust and from the Palestinian disaster. As Hana herself said: "Abed and myself, we are both children of refugees. My dad underwent traumatic suffering [during the Holocaust] and the 'evidence' [his home] was also erased" (Yaad, 24 December 2007). Some of the people even argued that it was precisely this common sense of refuge which enabled them to think about common future projects. Actually, the last meeting was dedicated to the question of future activities. The participants raised several ideas concerning the continuation of activities that included, among others: keeping personal contacts, organizing an annual meeting in the Miaar Hill to commemorate the Nakba and the Jewish memorial day, and organizing a new series of workshops, in order to enlarge the circle of participants on both sides. Another type of suggestion was to enable families of Miaar to live in Yaad[5] or alternatively to act together to promote the establishment of a new Arab village in the Galilee.[6] In the following paragraphs, I introduce in detail the two planning themes raised at the meetings as ideas for common projects, and the difficulties that arose in realizing some of the ideas for collaboration. As I argue later, it emphasizes how complex the process in the Galilee is and, therefore, how these negotiations should be investigated from the point of view of the role of conflicts in such dynamics, and how it is necessary to discuss these events using the terminologies of antagonism and agonism, rather than consensus and agreement.

"My Children are Very Attached to the Hill" (Jinjit, Yaad, 26 November 2004): The Development of the Miaar Hill

The first planning theme that was proposed during the workshops was the development of the Miaar Hill, where the ruins of the houses are still found. The idea suggested was to develop this site as a park, or as a prayer or a meeting place. It is interesting to note how emotional both sides felt towards this site. From the Palestinian side, this is quite obvious and clear: they feel strongly attached to this site because it represents their home. However, it is also highly important for the Yaad residents. One of the participants from Yaad said: "My children are very much attached to the hill, when they were young it was the site of our daily picnics after kindergarten. Here we had lunch and for them it's a home. Once we even found an ancient bone there. It was very tangible. We also found an ancient sarcophagus (burial coffin) and underneath, the remains of a Jewish village from the Roman period [AD 1–3]. I can imagine how the old Jews lived here in ancient times. I envy the Arab culture for their ability to pass their traditions to the children" (Jinjit, 26 November 2004).

Archeological history is used by the Jews to emphasize that the site offers many layers of memory and belonging and, in fact, that Miaar is only the last of

---

5   Yaad is a communal village, which means that new members can only be accepted following the approval of the general assembly.

6   Since 1948, no new Arab village has been established in the Galilee.

the historical villages to have been destroyed. On this historical basis, the idea of developing a park as a place of encounter between the three religions was raised, mainly by the Yaad residents. At the beginning, both sides agreed to initiate this project – they perceived it as a joint communal project that would connect the past to the present. Later on, the Palestinians changed their mind, claiming that perhaps they themselves would return and build their houses on the site, so that there was no logic on their part to want to build a park there. To quote Abed: "Why build a park? In whose memory? Memory parks don't give anything. The ruins should stay as they are so that other people know that there was a disaster in this village" (30 December 2009). Here is an example of an initial agreement that could not be realized because of different perceptions of remembrance, forgetfulness and belonging. In this case, there is no point in trying to reach an agreement, as the Therapeutic Planning approach aims to. Perhaps it was more appropriate to acknowledge the fact that the Hill is an agonistic space. As Connolly (1998) put it, these are political spaces that include a legitimate and public acknowledgment of the right to accessibility to resources. In this case, it is the right of the Palestinian past to be acknowledged.

"I Wish I Could Find One Courageous Jew Who Would Say Out Loud: 'They [the Palestinians] Deserve It [to Have What They Want]'" (Abed, Miaar, 15 October 2004)

The second planning theme that arose in the meetings concerned the rehabilitation and the fencing of the Palestinian graveyard. This issue was raised by the Palestinians. Initially, Hana and Abed wanted to promote this project but, as the meetings progressed, the issue raised a lot of objections from both sides. The Palestinians argued that they themselves were unable to promote the project without involving the Al Akza NGO. This NGO has been promoting the rehabilitation of Muslim graveyards in Israel since the 1980s (Fenster, 2004a). For that purpose Al Akza established an information network on all the graveyards in Israel, their location and areas. In another paper (Fenster, 2004a), I discussed in depth how the conflict developed in the early 1990s between Al Akza and the local Jewish council of Nesher, on the enlargement of the main road leading to the Nesher local municipality. This enlargement meant that some of the graves in the nearby Muslim graveyard would be destroyed. In the end, the parties involved reached a compromise, according to which the road would be built on columns, so that the graves underneath would not be harmed. In the case of Miaar, Abed did not want to pursue the project of the graveyard's rehabilitation without the involvement of Al Akza: "This is a religious matter and I am not an expert in that, I don't know … Al Akza people are the experts. They have the knowledge. They can get the maps of the graveyard before 1948 and can mark the borders of the graveyard … I can't work on this issue. It's a matter for religious people" (Abed, Miaar, 24 December 2009). Indeed, the Al Akza people arrived with maps of the area and marked the borders of the graveyard. This act, however, was what made Hana feel fearful and even panicky: "At the end, Al Akza people came, and a sheik (a leader) came with a beard[7] and another engineer full of good intentions and then I panicked, Who knows who these people are?" (Hana, 24 December 2007). She then turned to the person in charge of security in the region and told him about this meeting. His reaction was typical of the kind of mistrust and suspicion found on both sides. He warned her that "This is all evil and it's all negative". Hana then asked Abed to disconnect this project from Al Akza and to carry on with it by themselves, but Abed felt obliged towards them and did not feel that he could pursue the project locally without involving the relevant bodies, such as Al Akza. It seems that perhaps a "speak out" could have helped both sides understand the fears and hesitations of the other side, and perhaps open up the way to materialize this project. The graveyard is another example of an agonistic space, about which there is a competition as to recognition and preference, that leads to antagonism (Mouffe, 2002) and prevents the project from moving ahead.

"We Haven't Reached the Stage Where We Can Envision the Future" (Rachel, Yaad, 26 November 2004)

The difficulties involved in realizing the two planning projects emphasize how complicated it was for Yaad and Miaar residents to overcome past tragedies, memories and attachments, in order to meet present needs and future aspirations, by creating consensual daily spaces. While some of the Yaad residents wanted to concentrate on a vision of the future, rather than clinging to the past, others thought that it was impossible to envision the future without discussing the past, despite the difficulties. This was also what some of the Palestinians thought. However, how much of the past should be the focus of the discussions? The recent pre-1948 years? Or the more ancient times, when the area was populated by Jewish settlements (AD 1–3 centuries)? One of Yaad's residents thought that the main issue was what he termed "The wheel of history" or the layers of settlements in the area, referring mainly to the remains of the Jewish settlement dating from the Roman period, that were found near the ruins of Miaar. It seems like the archeological history of the site eased the sense of guilt felt by some of the Jewish residents, and put the 1948 tragic events into a different perspective. The Palestinians, on the other hand, did not like the idea that archeological history should become a topic of discussion. They took it as an escape from the real and painful problem: "You can't compare what happened thousands of years ago with what happened 60 years ago, when those who experienced the tragedy are still alive" (Wafik, Miaar, 30 December 2007). Despite these difficulties, the vision of the future and possible collaborations were discussed in the workshops. For example, despite their desire

---

7   Growing a beard is a religious practice among Jews and Muslims. Hana meant that this person was religious.

to return to Miaar, the Palestinians understood that they could not return to the same site, so one of the ideas they raised was to build a new settlement near Miaar, that would be either mixed Jewish and Arab, or just Arab. As I already mentioned earlier, this was a problematic suggestion, since no new Arab settlement has been established in the Galilee since 1948. But the dialectical feelings on the part of the Palestinians are part of these suggestions as well, because purchasing land from the Israel Land Authority means acknowledging its authority as landowner, while the Palestinians themselves claim to be the owners of the land in the area. That is why some of them do not see building a new settlement as a solution: "It is as if I admitted that this isn't my land and besides, we are willing to accept only a general solution, in relation to the entire Jewish–Arab conflict, together with the other 400 villages that were destroyed in 1948 (and not a local solution). What I want is that all the expropriated land be returned to us, that we have the right to build on our land. This is not possible at this stage" (Wafik, 30 December 2007). This quote emphasizes how complicated it is to let go of the past for the benefit of the present and the future. I would like to suggest that perhaps this situation of being stuck comes from the expectation that a consensus and an agreement can be reached, while it is perhaps more important to accept that there are various conflicts and disagreements and to be positive about the opportunities that present themselves to expose these disagreements. As Hillier (2003) notes, it is better to accept a situation of antagonism which arises from the conflict itself or, as she puts it, "to domesticate antagonism into agonism", i.e. to accept situations of conflicts as positive and constructive and to build a vision of the future on the basis of clear and transparent conflicts.

"Through My Own Pain I Could Connect to the Other's Pain" (Arik, Yaad, 31 December 2004): Concluding Comments

The story of Yaad and Miaar is the story of many other places in Israel and in other parts of the world. It is a story of one place that contains layers of contradictory memories and emotions. Such stories have been silenced along the 60 years of Israel's existence because of the goal of the Zionist national project to establish Jewish spaces throughout the State of Israel that, in fact, would replace the old pre-1948 Palestinian spaces. This project was very successful for many years but, as I showed at the beginning of the chapter, there is a current trend of change that mainly involves new and local dynamics that wish to remind, remember and commemorate these contradictory memories or, in Hillier's words, "to transform antagonism into agonism". Such a trend has begun in Yaad–Miaar and is taking place in other settlements in the Galilee. In the case of Yaad, the submission, in 2003, of the objections to the plan for expanding the village on the ruins of Miaar yielded a decision of the local committee of Yaad not to do so, despite the approval of the plan by the Regional Planning Authorities. This move was succeeded by a series of workshops with the residents of Yaad and Miaar. The emotional, mental and social processes that took place in these workshops yielded new insights and understandings among both parties. As the facilitators indicated, every party understood that none of them was right all of the time. Moreover, the residents of both sides learned to accept the fact that there was more than one truth and that it was legitimate for each side to demand that its truth be recognized by the other side. Another insight was the necessity of incorporating the past in future development plans. In both cases – the development of the Miaar Hill and that of the graveyard – past memories and a sense of attachment played major roles in the way development was perceived. As I mentioned earlier, these lines of thought mark new ways of approaching "planning" and "planning processes", which characterize post-modern thought. These dynamics in Yaad and Miaar can be discussed in light of approaches, such as Therapeutic Planning. Although this was not exactly the practice implemented there, it recalls the same dynamics, as it includes a process of recognition, learning and knowing that is typical of Therapeutic Planning. The process involving the recognition, learning and knowing of the "other" is one of the important dynamics that took place in these workshops. As Hana noted: "After the workshops, I understood that you can't heal the wounds without exposing them first. You can't bury it like the establishment has been doing along the years, as we didn't know about the Nakba in school or even in the army, and I became aware of this only when I turned 50. When you bury a wound it turns into a tumor. The workshops exposed the wounds to the sun" (24 December 2007).

Analysing these processes highlights the strengths and weaknesses of Therapeutic Planning. It suggests that in highly complex areas it is sometimes better to aspire for recognition, learning and knowing the conflicts and disagreements than trying to reach a consensus and agreement. It is, however, very clear that such painful places should be treated from a therapeutic perspective, while acknowledging that antagonism and disagreement are part of these processes. As Baum (1999) argues, the opportunity given to people to talk about their fears as part of a planning process opens up new opportunities for a better future. This kind of planning process, argues Baum, enables the change from a painful past to a better future.

References

Anderson, B. (1999) *Imagined Communities*, The Open University, Tel Aviv [Hebrew].

Babbitt, E., Steiner, P., Asaqla, J., Chomsky-Porat, C. and Kirshner, S. (2006) "Combining Empathy with Problem Solving: The Tamra Model of Facilitation in Israel", Report.

Baum, H. (1999) "Forgetting to Plan", *Journal of Planning Education and Research*, 19 (1): 2–14.

Bell, V. (1999) "Performativity and Belonging: An Introduction", *Theory, Culture and Society*, 16 (2): 1–10.

Ben Eliezer, U. (1999) "Is there a Civil Society in Israel? Politics and Identities in New NGOs", *Israeli Sociology*, 1, 51–99.

Crang, M. (1998) *Cultural Geography*, Routledge, London.

Connolly, W. (1998) "Beyond Good and Evil: The Ethical Sensibility of Michel Foucault", in Moss, J. (ed.) *The Later Foucault*, Sage, London, 108–128.

de Certeau, M. (1984) *The Practices of Everyday Life*, University of California Press, Berkeley.

Efrat, E. (1984) *Geography and Politics in Israel*, Achiasaf Publishers, Tel Aviv [Hebrew].

Falah, G. (1996) "Living Together Apart: Residential Segregation in Mixed Arab-Jewish Cities in Israel", *Urban Studies*, 33 (6): 823–857.

Fenster, T. (2002) "Planning as Control – Cultural and Gendered Manipulation and Mis-Use of Knowledge", *Hagar – International Social Science Review*, 3 (1): 67–84.

Fenster, T. (2004a) "Belonging, Memory and the Politics of Planning in Israel", *Social and Cultural Geography*, 5 (3): 403–417.

Fenster, T. (2004b) *The Global City and the Holy City: Narratives on Knowledge, Planning and Diversity*, Pearson, London.

Fenster, T. (2007) "Memory, Belonging and Urban Planning in Israel", *Theory and Criticism*, 30: 189–213.

Forester, J. (1999) *The Deliberate Practitioner*, The MIT Press, Cambridge, Massachusetts.

Foucault, M. (1982) "The Subject and Power" in Dreyfus, H. and Rabinow, P. (eds) *Michel Foucault: Beyond Structuralism and Hermeneutics*, Harvester, Brighton, 214–232.

Gidron, B., Katz, H. and Bar, M. (2001) "The Civil Society in Israel: Empirical Findings", Lecture given at the 4th annual conference of the Israeli Center of the Research of the Third Sector, Ben Gurion University, Beer Sheva.

Golan, A. (2002) "The Politics of Wartime Demolition and Human Landscape Transformations", *War in History*, 9 (4): 431–445.

Halbwachs, M. (1980) *The Collective Memory*, Harper & Row, New York.

Healey, P. (1997) *Collaborative Planning*, Macmillan, London.

Hillier, J. (1998) "Representation, Identity, and the Communicative Shaping of Place", in Leigh, A. and Smith, M.J. (eds) *The Production of Public Space*, Rowman & Littlefield, Oxford, 207–232.

Hillier, J. (2003) "Agonizing over Consensus: Why Habermasian Ideals Cannot be 'Real'", *Planning Theory*, 2: 37–59.

hooks, b. (1990) *Yearnings: Race, Gender and Cultural Politics*, South End Press, New York.

Ishai, Y. (1998) *Civil Society Towards the Year of 2000: Between Society and State*, The School for Social Work, Hebrew University, Jerusalem [Hebrew].

Ishai, Y. (2002) *Between Recruitment and Reconciliation*, Carmel Publishers, Jerusalem [Hebrew].

Jacobs, J. (1996) *Edge of Empire: Postcolonialism and the City*, Routledge, London.

Kalush, R. and Law Yone, H. (2000) "The National Home and the Personal Home: The Role of Public Housing in the Design of Urban Spaces", *Theory and Criticism*, 16: 153–180.

Khalidi, W. (1992) *All that Remains*, Institute for Palestine Studies, Beirut.

Kimmerling, B. (1977) "SoverEignty, Ownership and Presence in the Israeli-Palestinian Territorial Conflict: The Case of Ikrit and Bir'im", *Comparative Political Studies*, 10 (2): 155–176.

Kimmerling, B. (1983) *Zionism and Territory*, Institute of International Studies, Berkeley.

Leach, N. (2002) "Belonging: Towards a Theory of Identification with Space", in Hillier, J. and Rooksby, E. (eds) *Habitus: A Sense of Place*, Ashgate, Aldershot, 281–298.

Lefebvre, H. (1991a) *Critique of Everyday Life*, Verso, London.

Lefebvre, H. (1991b) *The Production of Space*, Blackwell, Oxford.

Mauffe, C. (2002) "Which Kind of Space for a Democratic Habitus?", in Hillier, J. and Rooksby, E. (eds) *Habitus: A Sense of Place*, Ashgate, Aldershot, 93–100.

Mitchell, D. (2000) *Cultural Geography*, Blackwell, Oxford.

Moris, B. (1997) *The Birth of the Palestinian Problem 1947–1949*, Am Oved, Tel Aviv.

Nitzan Shiftan, A. (2000) "Whitened Houses", *Theory and Criticism*, 16: 227–232 [Hebrew].

Nora, P. (1993) "Between Memory and History – On the Problem of Place", *Zmanim*, 12: 4–19 [Hebrew].

Ohana, D. and Westreich, R. (1996) "Introduction: The Presence of Myths in Judaism, Zionism and Israelism", in Ohana, D. and Westreich, R. (eds) *Myths and Memory in Israeli Awareness*, The Van Leer Institute and Hakibutz Hmeuhad Publishers, 11–40 [Hebrew].

Padan, Y. (2004) "Re-Placing Memory", in Yacobi, H. (ed.) *Constructing a Sense of Place*, Ashgate, Aldershot, 247–263.

Rabinowitz, D. (1997) *Overlooking Nazareth*, Cambridge University Press, Cambridge.

Rabinowitz, D. and Abu Baker, H. (2002) *The Stand Tall Generation: The Palestinian Citizens of Israel Today*, Keter Publishers, Tel Aviv [Hebrew].

Saadi, A. (2001) "The Withdrawal of the State and its Implications on the Palestinian Minority in Israel", in Peled, Y. and Ophir, A. (eds) *Israel: From a Recruited Society to Civil Society?*, The Van Leer Institute and Hakibutz Hamhuhad Publishers, Jerusalem and Tel Aviv, 337–349 [Hebrew].

Sandercock, L. (1998) *Towards Cosmopolis*, Wiley, London.

Sandercock, L. (2003) *Cosmopolis II: Mongrel Cities in the 21st Century*, Continuum, London.

Shenhav, Y. (2003) *The Arab-Jews: Nationality, Religion and Ethnicity*, Am Oved, Tel Aviv [Hebrew].

Sibley, D. (1998) "Problematizing Exclusion: Reflections on Space, Difference and Knowledge", *International Planning Studies*, 3 (1): 93–100.

Slymovics, S. (1998) *The Objective of Memory*, University of Pennsylvania Press, Philadelphia.

Yacobi, H. (2004) "Urban Iconoclasm: The Case of the 'Mixed City' of Lod", in Yacobi, H. (ed.) *Constructing a Sense of Place*, Ashgate, Aldershot, 165–191.

Yacobi, H. and Zfadia, E. (2004) "Territorial Identity among Immigrants in Lod", *Theory and Criticism*, 24: 45–72.

Yiftachel, O. (1998) "Planning and Social Control: Exploring the Dark Side", *Journal of Planning Literature*, 12 (4): 396–406.

Yiftachel, O. (2000) "Ethnocracy, Geography and Democracy: Notes on the Politics of Judaization of Israel", *Alpaim*, 19: 1–26 [Hebrew].

Yiftachel, O. (2001) "The Homeland and Nationalism", *Encyclopedia of Nationalism*, 1, 359–383.

Yona, Y. and Saporta, I. (2000) "Housing and Land Policy in Israel: The Limits of the Civil Discourse", *Theory and Criticism*, 16: 129–152.

Yuval Davis, N. (2003) "Belongings: In Between the Indegene and the Diasporic", in Ozkirimli, U. (ed.) *Nationalism in the 21st Century*, Macmillan, Basingstoke.

Zerubavel, Y. (1995) *Collective Memory and the Making of Israeli National Tradition*, University of Chicago Press, Chicago and London.

Zionist Archive (7/1/43) *El Maar*, IDF File 49/716.

Chapter 6

The Reconstructed City as Rhetorical Space: The Case of Volgograd

Elena Trubina

The legacies of war continue to affect cities long after the war has ended. During the war, some cities are destined to become places for collecting the nation's energies and defeating the foe. After the war, determined to rebound, they become reconstruction sites. One of these cities, Stalingrad, in the history of the twentieth century metonymically signifies a turning point in the war in Europe as well as the cruelty of contemporary warfare. Stalingrad (renamed Volgograd in 1961) keeps attracting visitors from many countries, especially Germany and Russia. What do these visitors see when they come? With its Stalinist architecture and the imprints of war, the city, for some visitors, seems like a time machine, and it is not by chance that the educated locals readily joke that, "We are still the most Soviet city that one can find in Russia". They mean not only that the city's urban fabric was almost entirely defined in the first decades after the Second World War, but the fact that, due to its role in Soviet history, the city has a considerable number of Stalin supporters. Many of its cultural practitioners see it as a site of patriotic memory of Soviet traditions, which seem to them to be the only ones available.

The key idea of recent collective, cultural and personal memory studies is that memory can no longer be seen as reflection, a transparent record of the past, but should be understood as essentially performative. Not only it is never morally or pragmatically neutral, but it can come into existence at a given time and place through specific kinds of memory work. In this chapter, I consider how in the city of Volgograd the built structure and the national traditions of mourning come into ambiguous interplay when seen from different points of view: that of the various generations of its inhabitants and visitors; that of the planners and developers who reconstructed it after the war; and that of the war memorials whose symbolic role and function have changed over time. Volgograd, while being an impressive example of socialist tradition in modernist planning, becomes for many the repository of "Soviet" memories both by virtue of its urban structure and the traumas of the Second World War. This is especially obvious in the everyday rituals through which memory work has been performed in Volgograd, a site sacred in Russian memory, and their reception by diverse audiences. In particular, I focus on the honour guard ceremony performed daily at the Eternal Fire memorial and on the rhetoric of excursion around the city.

Rhetorical Spaces

The complex relationships between the symbolic and the material are explored in the studies of visual rhetoric,[1] rhetoric's materiality,[2] rhetorical dimensions of urban sites,[3] and rhetorical space. It is the latter, used to describe the impact of physical space on a communicative event, which allows accounting for the sense of the city one gets as one joins a guided tour through present-day Volgograd. In Roxanne Mountford's article on the gender dimensions of church architecture, the rhetorical space of the pulpit is shown as embodying traces of history in the ways it physically represents relationships and ideas. Mountford emphasizes that "spaces have heuristic power over their inhabitants and spectators by forcing them to change both their behavior and, sometimes, their view of themselves".[4] For instance, those whose presence in such spaces is traditionally rendered problematic (women preachers in clerical settings) should not only re-imagine the space but symbolically trespass upon sacred ground. I'll apply this notion to the fleeting experience of taking part in an excursion through the city that still deals with its wartime traumas. This is an experience through which memory comes into existence at a given time and place for a particular audience. I believe it can be a useful way to deal with a location that seems to encourage one kind of utterance and performance and to discourage another. On the one hand, it is the variety of city sites and visual artefacts that together produce a set of imperatives defining what can be said about the city and how it should be seen. On the other hand, it is the predominant narratives that shape what we see or prompt us to see this or that site in a prescribed manner.

The circulation of the memory of the Second World War, both within and beyond the national narrative, today becomes more and more complex. On the one hand, people, especially the young and middle-aged, increasingly are exposed to diverse and contradictory impulses usually associated with globalization. On the other hand, the nation-state is determined to keep its hegemony of producing and reproducing the official narrative of that very particular historical event. The war in general and the Battle of Stalingrad in particular continue to be used as a source of legitimacy for the government and of pride for the populace, whose history is so problematic and troubled that victory in that war seems a single, untainted accomplishment. The politics of memory and commemoration are produced through the spaces of museums and memorials, the design of monuments, the contents of official speeches and excursions, and the meaning of public ceremonies.

---

1   Charles A. Hill and Marguerite Helmers, eds, *Defining Visual Rhetorics* (Mahwah, NJ: Lawrence Erlbaum, 2004).

2   Carole Blair, "Contemporary U.S. Memorial Sites", in *Rhetorical Bodies*, ed. Jack Selzer and Sharon Crowley (Madison: University of Wisconsin Press, 1999), 16–57.

3   Lawrence W. Rosenfield, "Central Park and the Celebration of Civic Virtue", in *American Rhetoric: Context and Criticism*, ed. Eugene E. White (University Park: Pennsylvania State University Press, 1980), 221–266.

4   Roxanne Mountford, "On Gender and Rhetorical Space", *Rhetoric Society Quarterly*, 31.1 (2001): 50.

Volgograd's case is special because of the way the city discursively renders the Battle of Stalingrad itself while telling the story of the city's reconstruction along with it. In one guidebook, for instance, it is put as follows: "The vicious enemy has destroyed all that was a subject of patriotic pride of the Soviet people, namely, the city plants that were created and reconstructed in the pre-war years. Of 126 plants and factories, none has survived ... The war was continuing but the life demanded to restore – under these complicated circumstances – the city industry, communal services, and housing. On the ruins the slogans emerged: 'We will reconstruct you, our beloved Stalingrad, from the warmth of burned-out areas and the ruins'".[5]

The reason for the continuous rhetorical intertwining of the event itself and the consequent reconstruction is that the battle resulted in nearly complete devastation, presenting postwar Soviet planners with a tabula rasa upon which they were expected to build an urban icon reflecting the invincibility of the Soviet regime. The predominant way of seeing the city can be found on the official site of the city. It displays a photograph of the city centred around the building in which a panorama of the Stalingrad battle is located and the House of Pavlov's ruins can be read first of all as an emblem of the city's resilience, its willingness to come back to life after being almost totally destroyed. It can be also read as a sign of the city authorities' continuing reliance on the memory of the Stalingrad battle in circumstances when the pressures and the consequences of deindustrialization become increasingly substantial. A formerly important industrial site, the city today is facing the prospect of becoming a "loser" city, and thus itself becoming a monument to Soviet planning with its focus on the needs of industry, putting the considerations of urbanity second. How, we may ask, have architecture and the very structure of the city been made to signify historical memory in Volgograd, and what have been the inadvertent effects of this signification?

The Memory of Volgograd's Reconstruction: Trauma and Resilience

The story of Volgograd's reconstruction is inseparable from the metanarrative of modernism with its faith in rational planning and the intertwining of the processes of industrialization and urbanization. The story of industrialization tells us that underlying the tendency of industrial enterprises to group together are the considerations of transactional efficiency and the realization of this efficiency by locating plants near sources of raw materials, power, transportation and labour. Cities are seen as the "natural" place to locate factories and firms. Thus, albeit in the first part of the twentieth century, urbanization, as the key to socialist strategy, was thought to serve the development of the productive forces. The socialist urbanization of Volgograd exemplifies a part of the Fordist urbanization:

---

5   Pyotr Gundyrin, *Puteshestvie po Volgogradu* (Volgograd: Nizhne-Volzhskoye knizhnoye izdatelstvo, 1978), 56–57 (translation by Elena Trubina).

functionalist industrial architecture, city-building industry, and modern urban planning – all these trends were mobilized to make the symbiosis of plant and neighbourhood efficient, inevitable and fitting to the inhabitants' subjectivities. Just like state socialist urbanization was only a version of modernist urbanization, so in the postwar urban reconstruction processes in Europe one can see common tendencies. Everywhere in Europe state planning reached its climax in the aftermath of the Second World War. To give just one example, the urban historians Daniele Voldman and Mark Clapson describe the postwar reconstruction process in France and Britain as one in which architects and planners became the major actors. "Prior to 1939, the planning profession had only a limited influence, but war extended and consolidated the influence of planners within the emergency system of 'experts' created by the coalition government", states Clapson.[6] Voldman emphasizes the blooming of the "technocratic structures" underlying the postwar planning that involved the expanded intervention of the state, the predominance of the idea of a managed economy, and a high degree of citizen support.[7]

The combination of the expanded intervention of the state and the high degree of citizen support that the French historian lists are two of the reasons why the period of intense reconstruction of industrial sites that immediately followed the Battle of Stalingrad and civic rebuilding continue to hold the imagination and pride both of urban inhabitants and planners in Volgograd and around Russia. Stalin said in his election speech of 9 February 1946, "The war was nothing other than a test of our Soviet system, of our state, our government, and our Communist Party".[8] The reconstruction process was also seen as a "test", which explains why in many accounts of this period the focus is on the leading part the Communist Party played in the development of government policy during this pivotal period in state-led planning, why there are few accounts that have examined the specific role of architects and planners in the planning and redevelopment of postwar Russian towns and cities using analyses of archival and published records, and why there is almost nothing devoted to the process of civic rebuilding. The most impressive way the local populace took part in Stalingrad's reconstruction was the so-called Cherkasovskoye movement, or Cherkasovskoye dvizhenie (under the name of the kindergarten nurse Alexandra Cherkasova) that began in 1943. Amidst the ruined city, 19 housewives and soldier's wives began looking for any houses suitable for living. They removed the corpses, collected useful items, and took care of many orphans, thus showing active concern for others. This voluntary movement has grown to a considerable size, and, in the recollections of the planners of the time, one can even find traces of a certain distress caused by this local initiative and the claims to social membership

6   Mark Clapson, *Invincible Green Suburbs, Brave New Towns* (Manchester: Manchester University Press, 1998), 38.

7   Daniele Voldman, *La reconstruction des villes francaises de 1940 a 1954. Histoire d'une politique* (Paris: L'Harmattan, 1977), 146.

8   Joseph Stalin, "Rech' na predvybomom sobranii ...", in *V. Stalin. Sochineniia*, 3 vols, ed. Robert H. McNeal (Stanford: Stanford University Press, 1967), Vol. 3, 4–5.

that it contained. The planners wanted to control everything in the process of the city's redevelopment while these initiatives from below, based on the desperate need to have habitable places, made the approaches to spatial solutions perhaps too divergent. In one of the short scholarly investigations of this important episode, the author emphasizes the persistence with which the authorities strove to keep this movement under control.[9] Recent books devoted to the postwar reconstruction rather positively depict the efficiency with which the centralized planned economy of the USSR could use all available resources in order to successfully rebuild the city in less than two decades.[10] At the same time, numerous deficiencies of postwar planning and building remain in many cases unanalysed.

The urban designer Kevin Lynch argues that the images of any given city that its citizens have should approximate the public image of the city or the set of public images each held by some group. When a subject finds herself operating within her environment, she organizes her city image from "five types of elements: paths, edges, districts, nodes, and landmarks".[11] Lynch argues that it is paths that are the most important parts of a city image: "People observe the city while moving through it, and along these paths the other environmental elements are arranged and related". Lynch means by "edges" edges of development that work as important organizing features for citizens because they play the role of "holding together generalized areas".[12] What comprises both the path and the edge in Volgograd's image is surely the river Volga. For ages, the Volga produced a surplus to support this city, to paraphrase Adam Smith. Its strategic location along a major transport artery enabled the city to become an industrial centre before the war. The river, as the city's supposedly "natural" feature, is in fact the reason why Volgograd is now counted among the longest cities in the world. In Soviet times, plant after plant after plant was built on its bank, and the city had to fill the spaces between the plants. The imprints of industrial activities that the city structure bears makes it a difficult setting for city life, especially from a transportation point of view.

When reconstruction started, the main task for urban planners was to figure out how Stalingrad could work through both the problematic legacy of its pre-war past and the massive destruction during the war. Before the war, the city comprised microraions surrounding the tractor plant, the metallurgical plant, the "Barricade" plant, and many others. Each plant had to be located as close to the river as possible. This resulted in what is called in Soviet planning jargon the "linear" or "stripe-

9   Elena Saharova, "Stranitsy istorii Cherkasovskogo dvizhenia", in *Stalingradskaya bitva v istorii Rossii. Vosmye yunocheskie chteniya. Sbornik dokladov*, ed. Michail Zagorulko (Volgograd: Volgogradskoye nauchnoe izdatelstvo, 2004), 192–201.

10   Julia Kosenkova, *Sovetsky gorod 1940-h-pervoi poloviny 1950-h godov. Ot tvorcheskih poiskov k praktike stroitelstva* (Moscow: Editorial URSS, 2000); Emma Kuzmina, *Vosstanovlenie Stalingrada. 1943–1950* (Volgograd: Izdatel, 2002).

11   Kevin Lynch, "The City Image and its Elements", in *The City Reader*, 2nd edition, ed. Richard T. LeGates and Frederic Stout (New York: Routledge, 2000), 479.

12   Ibid.

like" structure of the city. Plants were located along the river, one after the other, and urbanity had to follow these developments. There are many cities located on rivers, but there are few that wind up stretching 90 km along a riverbank. It is telling that in his essay, "Form of the Cities", Kevin Lynch cites Stalingrad as an example of the linear shape and argues as follows: "Overextension of the linear form risks throttling the main artery, particularly if too much local movement and access to the road is allowed. On a giant scale it cuts the countryside into the kind of isolated patches which now can be seen in the environs of big cities. There is an inherent lack of focus in this form, a lack of centers around which identification and activities can group themselves. Yet on a smaller scale or for particular uses this ancient form has value today."[13]

The contradiction underlying the process of subsuming the city's shape under the needs of a centralized economy was that the more industrial modernization developed, the more difficult it was to follow the ideal model of the socialist city in Soviet planning practice. In the Socialist ideal-typical model, the city was to be a combination of economic efficiency, social justice (in terms of access to urban goods and services), and a high quality of life for the urban population.[14] The model that was implemented in the Soviet Union had the following characteristics: industrialization and urbanization based on state ownership of the means of production and the centrally planned determination of the use and allocation of resources; priority given to investment and heavy industry; economic planning over physical (spatial) planning; investment-production plans and locational choices based not on market or profit criteria but on the planners' preferences, taking into consideration local, regional and national needs.[15] Stalingrad's reconstruction makes apparent the spatial contradictions of industrial settlements. However determined Volgograd's planners might have been to improve the quality of urban life (and thus to promote a way of living in line with socialist values) after the war, the pressures of the Soviet urban economy predetermined the reproduction of the pre-war linear plan. In capitalist countries after the end of the Second World War, according to Patrick Troy:[16]

> Industrial plants were released from the need to have rail sidings or a waterfront
> location and could locate to meet different site requirements ... The progressive
> move of industry from the older congested area to new locations on the fringes of
> the cities introduced new restructuring forces which were essentially centripetal

---

13   Kevin Lynch, "The Form of Cities", *Scientific America*, 190.4 (1954): 58.

14   David M. Smith, "The Socialist City", in *Cities After Socialism: Urban and Regional Change and Conflict in Post-Socialist Societies*, ed. Gregory Andrusz, Michael Harloe and Ivan Szelenyi (Oxford: Blackwell, 1996), 70–100.

15   Gregory Andrusz, "Structural Change and Boundary Instability", in *Cities After Socialism: Urban and Regional Change and Conflict in Post-Socialist Societies*, ed. Gregory Andrusz, Michael Harloe and Ivan Szelenyi (Oxford: Blackwell, 1996), 37.

16   Patrick Troy, "Urban Planning in the Late Twentieth Century", in *A Companion to the City*, ed. Gary Bridge and Sophie Watson (Oxford: Blackwell, 2003), 547.

in effect. That is, cities often had to make new investments in transport infrastructure, especially roads, to connect new industrial locations into the rest of the metropolitan area.

Due to the limited resources of postwar urban development in the Soviet Union, plants still had to have a waterfront location, on the one hand, while on the other hand, the political ambitions of the central authorities expressed themselves in the order that the planners received, namely, to design an imposing downtown waterfront along with many other public places. A city with a shortage of housing was forced to build a central waterfront with granite tribunes for the public to observe, supposedly, river parades and fireworks.[17] The everyday reality of living under the socialist spatial organization of population was thus subsumed under two major developments: the continuation of the forced modernization of industry by means of the reconstruction of old plants and the building of new ones; and the production of architectural iconicity and its relationship to the patriarchal socialist state. The monumental, symbolic and iconic dimensions of urban, socialist, central places kept reflecting the city-based rule of the Communist Party.

In Stalingrad-Volgograd, three meanings of reconstruction can be differentiated: a physical one, a reconstruction of the Soviet Union's national identity, and trauma recovery. While the first two meanings of reconstruction became combined in the predominant rhetoric, the third ended up being almost totally suppressed. While triumphalism pervaded the whole public space, while all city sites were mobilized to highlight patriotism, victory and courage, little was done to work through the traumatic dimension of the city's history during and after the war. The cynicism of the state propaganda expressed itself in the fact that the eyewitness accounts of the atrocities committed by the enemy, which were heavily promoted during the war to evoke affective response on the part of Soviet citizens, were shut down after the war was over as if the people's suffering did not matter unless it could be used against the enemy[18] or to serve as proof of the people's loyalty to the regime. Nowhere in the published accounts of the final years of the war and the first years after it can one find an attempt to theoretically assess the fact that the destruction of institutional and individual life in Stalingrad was so fundamental that no aspect of social or personal life was left unaffected. Both Western and Soviet authors seem to follow the progressivist narrative and emphasize the moral resilience, courage and fearlessness of the Russians. Together with historical accounts, the urban environment, too, is supposed to have a certain redemptive potential, so here arises a difficult task for an observer who finds herself being caught in a conflict of sorts between the empathy towards the citizens of the city and the need for a critical interpretation of the problematic ways of using the urban form and

---

17   The waterfront was supposed to bear Stalin's name but the city authorities gave up this idea after Stalin's death (when the construction was still on).

18   Evgenii Dobrenko, *Metafora vlasti* (Munich: Verlag Otto Sagner, 1993), 262–273.

fabric. In the reciprocal relationships between memory and both the city's iconic sites and its shape itself, it is difficult to discern where people's will to resilience ends and where trauma, unhealed and persistent, resides.

Young Honour Guards, Guides and their Audiences

What interests me in particular are the social dynamics and political repercussions underlying the activities of young people. What are the reasons that all the sanctimonious patriotic clichés and activities of the Soviet era, after being transmitted through high-school education, become so unreflectively reproduced? Whether one thinks of this phenomenon in terms of collective unconsciousness, social inertia, the conservatism of municipal authorities, or even genius loci, the nature of the peculiar imperatives the reconstructed city imposes on its inhabitants and the version of its identity it promotes appear worthy of reflection.

It is teenage girls who give some of the most peculiar performances that help sustain Volgograd's patriotic identity (see Figure 6.1). They take part in the daily honour guard ceremony at the Eternal Fire memorial, which makes one think not only about the implicit significance of gender relations for the construction of

Figure 6.1   Girls taking part in the daily honour guard ceremony at the Eternal Fire memorial

history and cultural memory but about the everyday life of the memorials. It is one thing to witness a parade or fireworks on Victory Day, with the subsequent emotional atmosphere, and quite another to be exposed to or to take part in a patriotic ceremony amidst mundane life. I was curious to know how the training of the guards is organized. It turns out that a special education centre exists in Volgograd whose aim is to promote the patriotic upbringing of Volgograd's youth. It is called Post Number One, referring to a particular episode of the Young Communist League's activities in the 1960s when the league's members in Volgograd were mobilized to be honour guards at the Eternal Fire in the Square of the Fallen Fighters. Since then, only the best high-school students receive the right to be guards. The schools take part in this ritual in turn. Still, I wanted to know whether there was a queue among those who wish to work as such, or is it an obligation? The girls on duty said they are not supposed to talk until they come back to school after performing the ritual. I remember how difficult it was to stay still for a half an hour while fulfilling my own guard duty (as a university student, by the Eternal Fire in my city) as well as that uneasy feeling of both reverence and doubt: if someone asks why I am here, what will I answer except that I was ordered to do so? But I belong to a different generation and see that times have changed radically. Why, now, do girls and not soldiers perform this honorary duty? Why make these 12-year-old girls repeat the steps we all know belong to military parades rather then let them enjoy the salsa? A performance that could draw attention to the repeated gestures of devotion and loyalty reveals that this repetition is not identical to its imaginary source, and that it generates reactions more ambivalent than it purports to induce.

In sketching the troubled history of modern social urban studies, the urban scholars Ryan Bishop and Gregory Clancey observe that "the City" acquired "a heroic status in both capitalist and socialist storytelling"[19] after the Second World War. One of the reasons for the theoretical neglect or silencing of urban catastrophes was the prevalence of the overarching progressivist narrative, according to which cities were supposed to be places of order, not catastrophe. This model of the city as hero was most impressively exemplified by the Russian government's deployment of the Battle of Stalingrad as a vehicle of socialist ideology. The victory-related triumphalism became the very foundation of the Soviet state.

After the Second World War, a pattern of narrative was established in which the city's history was subsumed under the monumental celebration of the nation-state on a global scale. An imposing commemoration site was built in Stalingrad in 1967, one of nearly 70,000 war monuments that had been built all over the USSR. Mass construction resulted in stylistic affinities: most memorials are hierarchical figural compositions surrounded by plates with names. By now, many of these monuments have lost their function as spaces of rhetorical power and only come

19   Ryan Bishop and Gregory Clancey, "The City as Target of Perpetuation and Death", in *Postcolonial Urbanism*, ed. Ryan Bishop, John Phillips and Wei Wei Yeo (New York: Routledge, 2003), 66.

to life, so to speak, during the mass commemoration ceremonies on Victory Day. Russian researchers of the memorials of the Second World War note that there is one universal obstacle anyone looking at these memorials in post-Soviet Russia would face: "Everybody has seen them and knows they exist, but there are few who could tell where exactly they are located".[20] On the other hand, since war memorials are often located in the central city, they become the locals' promenades, and their functions thus tend to shift towards quotidian public places.

There are a few memorial and monumental places in the city that are devoted to the Battle of Stalingrad: the Stalingrad Battle Panorama-Museum, the Pavlov House, and the Mamaev Hill memorial. The Panorama allows visitors to view the entire field of combat, the Pavlov House is a burnt-out shell of the building where Soviet soldiers defended themselves for two months, and the Mamaev Hill memorial was built 20 years after the battle. These historical monuments work as rhetorical artefacts that wouldn't serve their purpose without curatorial zeal. The city narrative that is performed during the guided tour is composed from the sites designed to educate, to honour, and to instil a feeling of unity with those who gave their lives. These are the places where people have come for generations to reflect on who we are and to remember the nation's sacrifices. The image of the city that the texts of the tour reproduce is one of a unique place with unique inhabitants that many enemies tried to conquer but always failed.

The victory becomes an emblem and a common signifier of the city's identity. This explains the immersion of both the city's authorities and locals in particular accounts of the past that have, since 1943, been invoked by the state as a way to strengthen national identity. In these accounts, a complete German defeat by the Soviet forces is interpreted as proof of the superiority of the Soviet way of life, the "failed prophecies" trope figures as rejoicing that Hitler's delusions proved wrong while the courage of Soviet soldiers and civilians was endless and awesome. The hundreds of thousands of casualties and the ruins of the city are turned into figures in that narrative as the price we had to pay for victory. The "failed prophecies" trope gets applied not only to the battle itself but to the reconstruction of the city. After the battle, so the narrative goes, visitors from the West advised the local authorities to give up the idea of rebuilding the city since it would be a lot cheaper to build a new city from scratch. But the local Communist Party committee and the people of Stalingrad were determined to bring the city back to life. The city was destroyed but we've rebuilt it. For a distant listener, it is almost impossible, in this version of the story, to disentangle where the benevolent influence of the Communist Party ends and where civic participation begins.

It appears that the narrative has not been revised since the late 1980s. The young tour guide, who readily reproduces all the rhetorical devices and the dominant narrative of Soviet times and enthusiastically reproaches those in the group who dare to express their doubts with regard to her words on the pretext

---

20   Natalya Konradova and Anna Ryleeva, "Geroi i Zhertvy. Memorialy Velikoi Otechestvennoi", *Neprikosnovenny Zapas*, 2–3 (2005), 139.

of "insufficient respect towards the memory of soldiers", makes one think of the difficulties of producing rhetorical means better suited to present-day audiences' needs and tastes than those worked out in Soviet times.

What I think differentiates the Soviet politics of memory is that the collective memory after 1945 was not divided into particular generations. The distance from the war increases, but the discursive renderings of those events that would be characteristic of a second, a third, and now a fourth generation haven't become prominent parts of the public discourse. There is a sense in which it is only when exposed to the versions of the past that too obviously belong to the past era that visitors to the city and its numerous patriotic sites begin to feel a discrepancy between the time they live in and the rhetoric that has been promoted. A striking example of this discrepancy and of what might be called the many-pacedness of historical change were the final words of my tour guide: "Together, we just have visited one of the greatest war memorials in the world. You see now just how big is our Motherland [she pointed her finger toward the main statue of Mamaev's Hill] and how small all of you are."

Nowhere do what John Gillis calls a state "bureaucracy of memory"[21] and John Bodnar the "dogmatic formalism"[22] of official memory find their expression more strikingly than in these narratives and the very itineraries of these tours. The city markets itself as predominantly a "hero-city". Both on the official site of the city and in the tourist agencies' brochures, it is named the "hero-city Volgograd" (and this is the title that the city, indeed, received from Soviet authorities). The legacy of the Battle of Stalingrad is considered not as a challenge to present-day Volgograd but as an important symbolic resource and a major tourist attraction. The city's marketing efforts thus nicely coincide with the conservative war memory discourse and the orthodox official interpretation of the Second World War. One might think that in order to attract today's more diverse and demanding audiences to the predominant narrative some others could have been added, given the obvious difference in meaning that this war has for different generations. However, not only is this not the case, but we see the young cultural practitioners eagerly participating in reproducing the old stories.

According to one rhetorician, "Displays are rhetorical because the meanings they manifest before situated audiences result from selective processes and, thus, constitute partial perspectives with political, social, or cultural implications".[23] In Volgograd, by virtue of its location amidst the so-called red belt of Russia, the conservative rhetorical space was created and, while excluding more problematic

---

21   John R. Gillis, Introduction to *Commemorations: The Politics of National Identity*, ed. John R. Gillis (Princeton: Princeton University Press, 1994), 3.

22   John Bodnar, "Public Memory in an American City: Commemoration in Cleveland", in *Commemorations: The Politics of National Identity*, ed. John R. Gillis (Princeton: Princeton University Press, 1994), 75.

23   Lawrence J. Prelli, Introduction to *Rhetorics of Display*, ed. Lawrence J. Prelli (Columbia, SC: University of South Carolina Press, 2006), 11.

discourses, especially those focused on traumatic city history, it encourages certain speakers and promotes conventional narratives.

Numerous reactions to the tours given in Volgograd can be found in the live journals of the young Russians who share their travel experiences on the Web. One of those who recently visited Volgograd writes as follows:

> The speech our guide delivered was monotonous both in terms of its inflections and its meaning. All its contents could be predicted from the onset since it only transmits the historical stereotypes everybody's familiar with: all German soldiers are cruel cowards, all Soviet soldiers are heroes who are eager to give their lives in the heroic acts, and all the battle is the conflict of good and evil. The Soviet soldiers thus are transformed into half-gods, totally different both from those who lived then and who live now. And war history, rather than being a realistic narrative, turns into mythology with the heroes carefully selected.[24]

It would be surely too naive to reproach numerous Volgograd tour guides for drawing this Manichean picture instead of providing the visitors with a more subtle and balanced version of the events in question. Yet we may better understand what caused such a reaction on the part of a visitor if we read the electronic guide the Stalingrad battle museum provides to its visitors. Here is a fragment of the Stalingrad Battle-Panorama Museum's advertisement: "Against the background of the military operations, the painters have resurrected the legendary heroic deeds of Stalingrad Battle – a signalman Matvei Putilov, who connected the ends of the torn wires in his mouth, a junior sergeant Nicholai Serdukov, who covered the embrasure of the enemy's gun with his chest, a pilot Viktor Rogalsky, who performed a air-ground ram attack, a nurse Anna Beschastnova, who saved the lives of thousands wounded soldiers by carrying them out of the battlefield, a Red Army soldier Michail Panikhaha, who jumped at the enemy's tank and caught himself on fire".[25]

The uneasiness this description causes has to do with the fact that today we value a single human life differently. Left to our private thoughts, we ask ourselves uncomfortable questions about what we would do under these circumstances and don't have answers. I believe that what makes the expression of the memory of the war in the city's space deficient is that the rhetorical space we find ourselves in is too heavily ritualized. It precludes discussion and seeks to impose on visitors a passive acceptance of the version of history that seeks to promote the smallness of each of us in the face of the larger-than-life preoccupations of the state. However,

---

24   Artyom Hitch-Hiker, "Po Dorogam RF. Volgograd", Chast 2. 30 October 2005. Message posted to: http://hitch-hiker.livejournal.com/30809.html (last accessed 22 September 2008) (translation by Elena Trubina).

25   Hudozhestvennaya Panorama, "Razgrom Nemetsko-Fashistskih Voisk pod Stalingradom". Retrieved 22 September 2008 from http://panorama.volgadmin.ru/panorama.html (translation by Elena Trubina).

what makes things even more complicated is that even if a charismatic leader of today decides to uncompromisingly disentangle the city's history and the deeds of his symbolic fathers, the very cultural history and tradition of imagery would preclude him from fully succeeding.

Volgograd's rhetorical spaces are paradoxical in many ways. Perhaps the most difficult paradox is that the unique national narrative of courage and resilience is bonded to a territory that remains a blood-drenched and bone-filled ground to an extent that exceeds human imagination because of the symbolic meaning its name had, not only for Stalin himself, but for many on the battlefield.

References

Andrusz, Gregory. "Structural Change and Boundary Instability". In *Cities After Socialism: Urban and Regional Change and Conflict in Post-Socialist Societies*, edited by Gregory Andrusz, Michael Harloe and Ivan Szelenyi, 30–69. Oxford: Blackwell, 1996.

Bishop, Ryan and Gregory Clancey. "The City as Target of Perpetuation and Death". In *Postcolonial Urbanism*, edited by Ryan Bishop, John Phillips and Wei Wei Yeo, 63–86. London and New York: Routledge, 2003.

Blair, Carole. "Contemporary U.S. Memorial Sites". In *Rhetorical Bodies*, edited by Jack Selzer and Sharon Crowley, 16–57. Madison: University of Wisconsin Press, 1999.

Bodnar, John. "Public Memory in an American City: Commemoration in Cleveland". In *Commemorations: The Politics of National Identity*, edited by John R. Gillis, 74–89. Princeton: Princeton University Press, 1994.

Clapson, Mark. *Invincible Green Suburbs, Brave New Towns*. Manchester: Manchester University Press, 1998.

Dobrenko, Evgenii. *Metafora vlasti*. Munich: Verlag Otto Sagner, 1993.

Gillis, John R. "Introduction". In *Commemorations: The Politics of National Identity*, edited by John R. Gillis, 3–26. Princeton: Princeton University Press, 1994.

Gundyrin, Pyotr. *Puteshestvie po Volgogradu*. Volgograd: Nizhne-Volzhskoye knizhnoye izdatelstvo, 1978.

Hill, Charles A. and Marguerite Helmers, eds. *Defining Visual Rhetorics*. Mahwah, NJ: Lawrence Erlbaum, 2004.

Konradova, Natalya and Anna Ryleeva. "Geroi i Zhertvy. Memorialy Velikoi Otechestvennoi". *Neprikosnovenny Zapas*, 2–3 (2005): 134–148.

Kosenkova, Julia. *Sovetsky gorod 1940-h-pervoi poloviny1950-h godov. Ot tvorcheskih poiskov k praktike stroitelstva*. Moscow: Editorial URSS, 2000.

Kuzmina, Emma. *Vosstanovlenie Stalingrada. 1943–1950*. Volgograd: Izdatel, 2002.

Lynch, Kevin. "The Form of Cities". *Scientific America*, 190.4 (1954): 53–64.

Lynch, Kevin. "The City Image and its Elements". In *The City Reader*, edited by Richard T. LeGates and Frederic Stout, 478–82. New York: Routledge, 2000.

Mountford, Roxanne. "On Gender and Rhetorical Space". *Rhetoric Society Quarterly*, 31.1 (2001): 41–71.

Prelli, Lawrence J. Introduction to *Rhetorics of Display*, edited by Lawrence J. Prelli, 1–38. Columbia, SC: University of South Carolina Press, 2006.

Rosenfield, Lawrence W. "Central Park and the Celebration of Civic Virtue". In *American Rhetoric: Context and Criticism*, edited by Eugene E. White, 221–266. University Park: Pennsylvania State University Press, 1980.

Saharova, Elena. "Stranitsy istorii Cherkasovskogo dvizhenia". In *Stalingradskaya Bitva v Istorii Rossii. Vosmye Yunocheskie Chteniya. Sbornik dokladov*, edited by Michail Zagorulko, 192–201. Volgograd: Volgogradskoye Nauchnoe Izdatelstvo, 2004.

Smith, David M. "The Socialist City". In *Cities After Socialism: Urban and Regional Change and Conflict in Post-Socialist Societies*, edited by Gregory Andrusz, Michael Harloe and Ivan Szelenyi, 70–100. Oxford: Blackwell, 1996.

Stalin, Joseph. "Rech' na predvybomom sobranii …". In *V. Stalin. Sochineniia*, 3 Vols, edited by Robert H. MacNeal. Vol. 3, 4–5. Stanford: Stanford University Press, 1967.

Troy, Patrick. "Urban Planning in the Late Twentieth Century". In *A Companion to the City*, edited by Gary Bridge and Sophie Watson, 543–554. Oxford: Blackwell, 2003.

Voldman, Daniele. *La reconstruction des villes francaises de 1940 a 1954. Histoire d'une politique*. Paris: L'Harmattan, 1977.

Chapter 7

Seoul: City, Identity and the Construction of the Past[*]

Guy Podoler

Introduction

Seoul was designated the capital city of Korea in 1394, with both strategic and cultural-related considerations determining its selection and layout. Nestling at the foot of mountains, the traditional city was also protected by the Han River to the south and by a wall that surrounded it. Under Japanese colonial rule (1910–45) the city markedly changed as construction and demolition projects transformed it into a modern centre with a population of about one million. In 1948, with the formal split of the Korean nation, Seoul was named the capital city of South Korea, and except for a tragic temporary setback during the Korean War (1950–53) it has continuously been expanding since. In the 1960s it started to spread south of the Han River and today it is among the most populous mega-cities in the world, with over 10 million inhabitants – nearly a quarter of the country's total population.

Woven into the process of Seoul's postcolonial growth was the attempt to define and redefine South Korea's national identity through intentional changes in the urban landscape. Accordingly, this chapter aims to highlight some of these familiar and well-publicized changes and analyse them in order to present a case study, which is intended to contribute to our understanding of the role of the urban landscape in mirroring and shaping national identity. While recent works have problematized the complex connection between space and Seoul's "city identity" by exploring, among others, the existence of the city's foreign communities, the differences between its consumer spaces, and the contested nature of its club culture,[1] this chapter focuses on the relationship between space, identity and

---

  *   This chapter is a modified version of my Hebrew-language article "Se'ol: Hebetim shel havnayat zehut be-emtsa'ut ha-nof ha-ironi", in *Zikaron, Hashkahah, ve-Havnayat ha-Merhav*, ed. Haim Yacobi and Tovi Fenster (Jerusalem: Hakibbutz Hameuchad and Van Leer Institute, forthcoming) (Hebrew).

    1   See, respectively: Eun Mee Kim and Jean S. Kang, "Seoul as a Global City with Ethnic Villages", *Korea Journal* 47, no. 4 (Winter 2007): 64–99; Dong Yeun Lee, "Consuming Spaces in the Global Era: Distinctions between Consumer Spaces in Seoul", *Korea Journal* 44, no. 3 (Autumn 2004): 108–137; and Mu-Yong Lee, "The Landscape of Club Culture and Identity Politics: Focusing on the Club Culture in the Hongdae Area of Seoul", *Korea Journal* 44, no. 3 (Autumn 2004): 65–107.

memory. My main concern is to explore the contested terrain of identity formation by connecting the city's developmental history with its postcolonial history of establishing, demolishing, restoring and arguing about prominent colonial-related "mnemonic" and historical sites.

The chapter begins by presenting the theoretical considerations that guide my analysis of the nexus between urban development and design, and identity construction. Next is a survey of Seoul's historical development, followed by a section dedicated to the formation of postcolonial identity through the city's urban landscape. In this regard, I focus on the "construction of memory" through museums, monuments and parks. Finally, a concluding section is presented.

Myth and the City

The urban landscape, as Tim Hall reminds us, is produced, regulated and consumed.[2] Deriving from this observation is the dynamic characteristic of the physical components of the city in terms of, first, their construction, renovation and/or demolition, and second, the significances and meanings inscribed to them by both their planners and their target audiences. "Cities contain multiple landscapes", explains Douglass, as "no single interpretation or vision of the city and its constituent elements is totalizing".[3] Accordingly, enveloped in the city's layout, buildings, parks, monuments, etc., is a rich terrain of human interests and complicated relationships of both concurrence and contention. In order to unpack a representative portion of these dynamics and to link it to the dynamics of shaping South Korean identity in Seoul I draw from, and connect between, two models. The first is the global-local framework as expounded in the works of sociologists Richard Child Hill and Kuniko Fujita[4] and urban planning specialist Mike Douglass,[5] and the second is the memory-myth dichotomy as presented by political theorist Duncan S.A. Bell.[6]

In a recent critique on the thesis of the overarching and converging characteristic of globalization and its effect on the development of cities – especially as presented

2   Tim Hall, *Urban Geography*, 3rd edn (London and New York: Routledge, 2006), 15–16.

3   Mike Douglass, "Urbanization and Social Transformations in East Asia", in *Culture and the City in East Asia*, ed. Won Bae Kim, Mike Douglass, Sang-Chuel Choe and Kong Chong Ho (Oxford: Clarendon Press, 1997), 44.

4   Richard Child Hill and Kuniko Fujita, "The Nested City: Introduction", *Urban Studies* 40, no. 2 (2003): 207–217.

5   Douglass, "Urbanization and Social Transformations", 41–65. Although the term "global-local framework" appears in Douglass' work but is absent from Hill's and Fujita's, the latter's view of "nested cities" is compatible with the framework employed by the former.

6   Duncan S.A. Bell, "Mythscapes: Memory, Mythology, and National Identity", *British Journal of Sociology* 54, no. 1 (2003): 63–81.

by Saskia Sassen's noted yet controversial work[7] – Hill and Fujita proffer the concept of nested cities. Drawing from theorists of the social sciences, they explain that, instead of thinking of hierarchy merely as a top-down relationship, hierarchy is "the nesting of parts within larger wholes, the relationship of levels", which means that "everything is both part and whole, both cause and effect, both separate from and dependent upon other entities".[8] According to this view, the world's major cities are nested in multilevel configurations – namely, "global niche, regional formation, national development model, [and] local historical context" – hence they evolve in various trajectories.[9] Thus, as Douglass puts it, a global-local framework "accepts a more open-ended, historically contingent process of development" as compared with modernist and Marxist ideologies, as well as with the idea that there is a unique Asian way of development that all, and only, Asian societies go through.[10]

Although Douglass also maintains that powerful external forces are "filtered through historically differentiated constellations of socio-cultural, economic, and political structures",[11] Hill and Fujita assert that "nation-states and cities do not simply 'filter' or 'mediate' global functions".[12] The latter observation, I believe, is more helpful in realizing the dynamics involved, since "causality runs back and forth among levels of society" and "interaction creates new configurations".[13] In any case, the point to be stressed is the existence of interplay between socio-cultural developments and the city, interplay that both results from and creates uniqueness.[14]

Under this context, if we employ the term "culture" to signify "a set of values, attitudes, and institutions that shapes social action just as action in turn shapes cultural change", then the city is "the spatial correlate of culture"[15] in a sense

---

7   Saskia Sassen, *The Global City: New York, London, Tokyo* (New Jersey: Princeton University Press, 1991).

8   Hill and Fujita, "The Nested City", 207–208.

9   Ibid., 212.

10   Douglass, "Urbanization and Social Transformations", 43–44. It is interesting to note that, in their work on the cities of New York, Tokyo and Seoul, Hill and Kim coupled the latter two to challenge the "world city paradigm" as advanced by Saskia Sassen and John Friedman (see Richard Child Hill and June Woo Kim, "Global Cities and Developmental States: New York, Tokyo and Seoul", *Urban Studies* 37, no. 12 (2000): 2167–2195). Drawing from the heated debate that ensued, Wang finds Hill's and Kim's arguments more compelling, yet given the findings of his own empirical work on another East Asian city, Taipei, he calls for further sophistication of the "Hill-Kim binary framework" (see Chia-Huang Wang, "Taipei as a Global City: A Theoretical and Empirical Examination", *Urban Studies* 40, no. 2 (2003): 309–334).

11   Douglass, "Urbanization and Social Transformations", 42–43.

12   Hill and Fujita, "The Nested City", 213.

13   Ibid.

14   This theme is at the heart of Won Bae Kim et al., eds, *Culture and the City in East Asia* (Oxford: Clarendon Press, 1997) from which I also adopt Douglass' analysis of the global-local framework.

15   Won Bae Kim, Mike Douglass and Sang-Cheul Choe, "Introduction", in ibid., 3.

that, the city is at the same time an outcome, a reflection and a producer of socio-cultural conditions and changes.[16] Deeply embedded in this dynamic process is the act of defining, redefining and/or sustaining identities on private, collective and national levels. Architect and theorist Neil Leach likened identity to a filmscript and architecture to the filmset, which "derives its meaning from the activities that have taken place there". "Memories of associated activities", he adds, "haunt architecture like a ghost";[17] yet for analytic purposes, "memories" and their connection to "identity" require further conceptualization.

Group identity will be understood here as "a sense of sameness over time and space", which is politically and socially constructed.[18] Seen in this light, a commonly held perception is that memory and identity engage in reciprocal relations since memory supplies identity with meaning while the certain identity defines what is remembered.[19] However, following Bell, in order to decipher the meanings imprinted in national identity and reveal the dynamics involved in establishing them, memory should be distinguished from mythology.[20] Memory is a "socially-framed property of individual minds" that may be collectively acted out through, for example, commemorative rituals, thus it "is not transferable (as memory) to those who have not experienced the events".[21] Myth, on the other hand, is transferable. Myth is a story, a narrative that "serves to flatten the complexity, the nuance, the performative contradictions of human history; it presents instead a simplistic and often uni-vocal story".[22] And in this regard, nationalist governing mythology, or the dominant narrative, seeks to "impose memory" – it "attempt[s] to impose a definite meaning on the past, on the nation and its history".[23]

---

16   Douglass, "Urbanization and Social Transformations", 45. Douglass draws here from various scholars whose observations I found best to summarize in this three-layered characterization.

17   Neil Leach, "Belonging: Towards a Theory of Identification with Space", in *Habitus: A Sense of Place*, 2nd edn, ed. Jean Hillier and Emma Rooksby (Aldershot: Ashgate, 2005), 308.

18   John R. Gillis, "Memory and Identity: The History of a Relationship", in *Commemorations: The Politics of National Identity*, ed. John R. Gillis (New Jersey: Princeton University Press, 1994), 3–5.

19   Ibid., 3.

20   It is beyond the scope of this chapter to delve into the voluminous body of literature associated with the concept of memory in its collective manifestation. A good survey is Jeffrey K. Olick and Joyce Robbins, "Social Memory Studies: From 'Collective Memory' to the Historical Sociology of Mnemonic Practices", *Annual Review of Sociology* 24 (1998): 105–140.

21   Bell, "Mythscapes", 72–73.

22   Ibid., 75. Though not explicitly mentioned by Bell, the function of "myth" as presenting a natural image of a historical reality through symbols is central to Roland Barthes' detailed analysis in *Mythologies*, trans. Annette Lavers (New York: Hill and Wang, 1972), 109–159.

23   Bell, "Mythscapes", 74.

The multiple landscapes of the city, then, are created through social change and are manifestations of relations existing among decision-makers, capital owners, planners and social actors. Since in many cases they are designed, among others, to make a statement about, and/or sustain, particular socio-cultural conditions, they are contested grounds where myths and memories are employed to define and redefine identities. It should be stressed in this regard that, just as some places, either within or without the city, are actively "forgotten",[24] "forgetting" is also intentional in the tangible construction of identity. In the section below I lay the historical backbone for the section that then follows, and that demonstrates how changing socio-cultural conditions, urban landscapes and shaping identity have been linked together in postcolonial Seoul.

Seoul's Historical Development: An Overview

In 1394, two years after founding the Chosŏn dynasty (1392–1910), General Yi Sŏng-gye designated Seoul, which was then called Hanyang, as his capital city. The city was chosen for both its strategic and geomantic advantages. Namely, the four mountains to the north and the Han River to the south provided natural defences with a concomitant harmonious surrounding, which on the basis of the "wind and water" theory (*p'ungsu* in Korean; *feng-shui* in Chinese) was perceived as a supplier of power and positive energy to the city.

Tradition and practicality also converged to determine the design of the new and well-planned capital. For example, the wall that was built around the city for protection had eight gates, four of which were main gates facing the cardinal points and each gate assigned to serve a different category of visitors. Within the walls the king's palace was constructed at the northern part of the city, protected by the mountains and facing the "good" direction of the south. Altars for the gods of earth and crops, and Chong'myo – the royal shrine where the kings performed their official duty of honouring their ancestors, and where the royal ancestral tablets were (and still are) housed – were built and positioned in a careful manner with relation to the royal palace to support the power of the monarchy and the dynasty.[25] Also, the area of residence and the size of the land and the house were determined by social class, thus, influenced by the dominating Confucian ideology, the physical layout of Seoul was "an embodiment of social

---

24   For a recent work see Yong-Sook Lee and Brenda S.A. Yeoh, eds, *Globalisation and the Politics of Forgetting* (London and New York: Routledge, 2006).

25   Un Rii Hae, "Jongmyo (Royal Shrine): Iconography of Korea" (paper prepared for the ICOMOS 14th General Assembly and Scientific Symposium: Place—Memory—Meaning: Preserving Intangible Values in Monuments and Sites, Victoria Falls, Zimbabwe, 27–31 October 2003), www.international.icomos.org/victoriafalls2003/papers/A2-7-%20UnRii.pdf (accessed 27 May 2008).

hierarchies".[26] "From this time on", as the historian Lee Ki-baik points out, "Seoul has been the political, economic, and cultural center of Korea".[27]

In the late sixteenth and early seventeenth centuries Chosŏn Korea suffered from foreign invasions that brought much destruction to Seoul. As a result, not only was the capital twice abandoned by the kings and their close aides who fled to find refuge, but much of the city's original form was devastated. During the decades that followed, Seoul was gradually rehabilitated and again prospered. Its official boundary, it must be noted, was not limited to the original 16 square kilometre walled area, but also included the rural suburbs that developed outside around the wall.[28] And with the ruling elite constituting in the mid-seventeenth century about 10 per cent of the population, it is possible to describe the city as "a political machine used by a ruling class whose economic basis of power lay in the agricultural production of the countryside".[29] Finally, by the late Chosŏn dynasty, Seoul's layout of "multiple landscapes" was a result of important socio-economic and socio-political developments; namely, the shortage of land, which caused both further divisions of already allocated units and illegal expansions, and disruptions to the strict class system, which saw the construction of disorganized structures.[30]

During the latter half of the nineteenth century, China (Korea's traditional patron), Western powers and Japan were vying with each other over influence on Chosŏn Korea, each striving to secure its economic, political and strategic interests on the peninsula. The emergence of Japan as the triumphant party from these struggles peaked in 1910 with the annexation of Korea, and with it came a decisive period in modern Korean history. Seoul – now called by the Japanese name of "Keijō" – was redesigned in accordance with the colonizer's interests, which to a large extent were to exploit and to impart colonial domination. Thus, as Pai has shown, modern urban planning transformed the city through new patterns of land division and the construction of new infrastructure, factories and modernly (Western) styled buildings. The walls surrounding the traditional city were torn down, and Seoul became a bustling centre of commercialism, consumerism, banking and entertainment. The structure of colonial Seoul conveyed that the old regime was gone and that a new and powerful one now governed.[31]

---

26 Hyungmin Pai, "Modernism, Development, and the Transformation of Seoul: A Study of the Development of Sae'oon Sang'ga and Yoido", in Won Bae Kim et al., eds, *Culture and the City in East Asia* (Oxford: Clarendon Press, 1997), 108.

27 Ki-baik Lee, *A New History of Korea*, trans. Edward W. Wagner with Edward J. Shultz (Cambridge, MA: Harvard University Press, 1984), 165.

28 Won-Yong Kwon and Kwang-Joong Kim, "Introduction", in *Urban Management in Seoul: Policy Issues and Responses*, ed. Won-Yong Kwon and Kwang-Joong Kim (Seoul: Seoul Development Institute, 2001), 3–4.

29 Pai, "Modernism, Development", 107.

30 Ibid., 108–109.

31 Ibid., 109–115.

At the aftermath of liberation in 1945 Seoul was in complete disorder as its Japanese residents, who constituted about 20 per cent of the population, had left, and as Korean refugees from Japan, Manchuria and the northern part of the peninsula poured in. In 1948 Korea was officially divided with the establishment of two separate states and Seoul became the capital city of the south.[32] During the Korean War (1950–3), which perpetuated the division, the city suffered massive destruction as about 30 per cent of homes and 70 per cent of factories and other public facilities were destroyed.[33] In the 1950s, during the presidency of Syngman Rhee (1948–60), there was no "systematic intervention" in Seoul[34] (see next section), and it was under the authoritarian rule of former general Park Chung-hee, who governed the country between 1961 and 1979, that the city was drastically redeveloped.

Park, "the builder of modern Korea", initiated consecutive Five-Year Economic Plans to rehabilitate and strengthen the country, and especially the second plan, launched in 1966, had a profound effect on the capital. Seoul mayor at that time was Kim Hyun-ok – a former military man whom the president had trusted, and who lived up to his nickname "bulldozer" by pushing forward major projects such as Central Business District (CBD) redevelopment, central throughways and more.[35] As a result of the massive development, the period witnessed the highest rate of increase in road construction in the city's history: while between 1960 and 2000 the total length of roads increased to 5.9 times the original length, between 1965 and 1970 alone it increased by 3.7 times.[36] Furthermore, while in 1966 almost 3.8 million residents lived in Seoul – forming 13 per cent of the south's total population – in 1970 the number grew to approximately 5.4 million, or 17.6 per cent of the population.[37] This, in turn, encouraged the authorities to launch a greenbelt strategy intended to limit the city's physical growth in 1971.[38] In 1973 Seoul's administrative boundaries reached 605 square kilometres, which is about their size today.[39]

---

32   In the final analysis, although local social and political forces of diverse colors and shades existed in liberated Korea, it was the power struggle between the Americans and the Soviets between 1945 and 1948 that eventually led to the establishment of two separate regimes. See Adrian Buzo, *The Making of Modern Korea* (London and New York: Routledge, 2002), 50–70.

33   Tong-Hyung Kim, "Seoul Sees Rapid Growth as Global Metropolis", *Korea Times*, 15 August 2005.

34   Pai, "Modernism, Development", 115.

35   Ibid.

36   Seoul Development Institute, "Infrastructure", in *Changing Profile of Seoul: Major Statistics and Trends* (Seoul: Seoul Development Institute, 2005), 82. Available at www.sdi. re.kr/nfile/about_seoul/contents/profile_seoul_4.pdf (accessed 28 May 2008).

37   Kunhyuck Ahn and Yeong-Te Ohn, "Metropolitan Growth Management Policies in Seoul: A Critical Review", in *Urban Management in Seoul*, ed. Won-Yong Kwon and Kwang-Joong Kim (Seoul: Seoul Development Institute, 2001), 53.

38   Ibid., 57.

39   Kim, "Seoul Sees Rapid Growth".

More construction works were carried out in the 1980s to meet contemporaneous conditions. Under the authoritarian government of President Chun Doo-hwan (1980–8) these conditions included, among others, a growing middle class – which was the fruit of economic growth – and Seoul winning the bid for the 1988 Olympic Games. Accordingly, massive housing projects for the middle class were built along with various facilities required for hosting the Olympics, and three more subway lines opened in addition to Line 1 that was opened in 1974.[40] In 1989 the government announced its decision to build five satellite cities outside the greenbelt in light of the economic boom that had led to a significant increase in the demand for houses, and, as a result, to an inflation in housing prices.[41] Seoul's population during that decade grew from just over eight million in 1979 to over 10 million for the first time in 1988.

Throughout the 1990s, Seoul's population fluctuated since marking a record number of almost 10.97 million residents in 1992 and until it has stabilized at around 10.3 million since 2001.[42] One main reason for this trend is the aforementioned satellite cities that have absorbed well over a million residents. Furthermore, the industrial structure of the city has changed with the growth of the high-tech, information and communication industries. In particular, Seoul's industries remarkably transformed as a result of the grave financial crisis of 1997, the crisis which left South Korea no choice but to receive a substantial sum of loans from the International Monetary Fund (IMF).[43] Then, only competitive firms could survive and prosper, and most of the new "venture firms" that have mushroomed in Seoul by the early 2000s, have been information processing and other computer related industries.[44]

In this regard, the darker side of South Korea's recovery from the 1997 crisis manifested itself in Seoul with the exacerbation of a pre-existing condition of inequality. As Yim has shown, the high-tech industries are concentrated in a specific area of the capital, an area that the authorities have been purposely privileging for years through, for example, the structure of the subway lines, the transfer of elite high schools from other areas and the establishment of theatres, museums, galleries, etc. This southeastern area of Seoul is thus recognized as the

---

40   Ibid.

41   Ahn and Ohn, "Metropolitan Growth Management Policies", 63.

42   Seoul Metropolitan Government Website, http://english.seoul.go.kr/index.html. Population statistics on this site are available through this link: http://english.seoul.go.kr/today/about/about_12stat.htm (accessed 28 May 2008).

43   The Asia Financial Crisis, or the "IMF Crisis", hit the economies of several Southeast and East Asian countries. In the end of 1997 an unprecedented IMF bailout package for South Korea, totaling over 58 billion dollars, was announced; yet already in December 1999 the South Korean president announced the end of the crisis, and by the summer of 2001 the country repaid the 19.5 billion dollar loan it had received.

44   Chang-Ho Shin and Chang-Heum Byeon, "New Industrialization in Seoul: Industrial Restructuring and Strategic Responses", in *Urban Management in Seoul*, ed. Won-Yong Kwon and Kwang-Joong Kim (Seoul: Seoul Development Institute), 131.

residency of the high-income population where land and apartment prices are the highest in the city. In short, Seoul has been witnessing what Yim calls a "process of socio-spatial polarization".[45]

In conclusion, power relations manifested themselves through Seoul's urban landscape since traditional times. As a general survey, this section did not elaborate on the intricate "nestedness" of these relations; instead, it focused on built icons of political domination, as well as on top-down plans concerning residential and developmental areas that have anchored socio-political relationships under changing conditions. In the first decade of the twenty-first century, two well-publicized projects initiated by Seoul Metropolitan Government sparked heated public debates, thus underscoring the controversial feature embedded in such significant alterations of the city's landscape. One initiative was the restoration of the Chŏnggye stream (Chŏnggyechŏn) (2003–5), and the second was the demolition of Tongdaemun Stadium (2008) in favour of a planned "World Design Park and Plaza". In both cases critics raised their concerns, among others, over what they viewed as an improper and insensitive treatment of the city's history.[46] In what follows, I attempt to shed light on how the fact that political and social actors in the postcolonial period have regarded it as most important to define what should be remembered, has manifested itself through the development of "Seoul's landscape of myth and memory".

Myth, Memory and Seoul's Memorial Sites

Concrete memorial sites constitute a particular powerful form by which identity is constructed by representing a past and conveying it to target audiences through myths and/or by invoking memories. The strength of these sites lies in their immediate (strong) visual effect and in their availability to people from all walks of life. The symbolism of a memorial, however, is not bound to remain static. A memorial may lose its appeal at some point in time, it may regain significance at a different time or it may not even have any appeal or significance to begin with. Also,

---

45   Seok-Hoi Yim, "Geographical Features of Social Polarization in Seoul, South Korea", in *Representing Local Places and Raising Voices from Below*, ed. Toshio Mizuuchi (Osaka: Osaka City University, 2003), 39. Available through this link: www.lit.osaka-cu.ac.jp/geo/e-frombelow.htm (accessed 27 May 2008).

46   On Chŏnggyechŏn and the debates surrounding it see, e.g., "Mayor Lee's Surprise Remark – 'Will Use City Budget for Anti-Capital Move' – Lee Fights Back Against Ruling Uri Party Attack", *The Seoul Times*, n.d. (2004?) (available at http://theseoultimes.com/ST/?url=/ST/db/read.php?idx=1049; accessed 27 May 2008); Jin-woo Lee, "Bribery Scandal Mars Restoration Project", *Korea Times*, 11 May 2005; and, Soo-mee Park, "Controversy Continues Over Chosen Cheonggye Sculpture", *JoongAng Ilbo*, 17 February 2006. On the controversies related to the Tongdaemun Stadium demolition see Soo-mee Park, "Local Architects Lash Out as Bulldozers Raze Dongdaemun", *JoongAng Ilbo*, 25 January 2008 (available at http://joongangdaily.joins.com/article/view.asp?aid=2885521; accessed 27 May 2008).

images and messages conveyed by memorials may be altered by the introduction of new meanings. Seen in this light, the demolishing of a monument or a memorial site is no less significant than the building of one. In order to demonstrate the dynamic role of Seoul's "mnemonic sites" in the attempts to mirror and ground national identity, I divide the turbulent history of South Korea into three parts: before, during and following the Park Chung-hee era.

The First Republic under President Syngman Rhee, 1948–60

In more than one way this was a time of uncertainty for the new country. South Koreans were then adjusting themselves to living under the conditions of the tragic division, and this was especially meaningful at the aftermath of the devastation left by the Korean War (1950–53). The country was destitute – in 1960 its economic structure and output resembled that of the peninsula's southern (colonial) provinces in 1940[47] – and although it was not the only reason for the country's inability to recover in the 1950s, President Rhee's administrative and managerial weaknesses played a decisive role in this failure.[48] It was also a period of urbanization and booms in education and in the press,[49] on the one hand, and a strongly contrasting rural social system,[50] on the other hand. As for Rhee's style of government, despite earlier expectations for a liberal democracy, the president established a corrupt autocracy, which was finally ousted by an urban civilian uprising on 19 April 1960.

With regard to the relationship with Japan, memories of the colonial period were still very much fresh and the president displayed animosity in his diplomatic engagements with the former colonizer. Next to his strong anti-communist rhetoric aimed to counter and de-legitimize internal opposition and dissent, Rhee employed harsh anti-Japanese rhetoric as well. One thus might have expected that Rhee, a former independence activist, would have capitalized on the memories of the colonial past to strengthen his regime's legitimacy. However, this was not the case for this past was problematic for the president from two aspects: first, Rhee's administration was stained by a collaborationist blot – it relied heavily on Koreans who had previously been employed by colonial authorities;[51] and second, Rhee's history included personal rivalries and schisms with other independence fighters. These two conditions significantly contributed, in my view, to Rhee's reluctance to fully engage in acts of "mythesizing" colonial memories, acts that might have drawn attention to the regime's uncomfortable connection with what was then a very recent past.

---

47 Buzo, *The Making of Modern Korea*, 104.
48 Ibid., 106; Mark L. Clifford, *Troubled Tiger: Businessmen, Bureaucrats, and Generals in South Korea*, rev. edn (New York: M.E. Sharpe, 1988), 30.
49 Eckert et al., *Korea Old and New: A History* (Seoul: Ilchokak, 1990), 353.
50 Buzo, *The Making of Modern Korea*, 102.
51 For example, about 85 per cent of Koreans who had served in the Japanese police force were employed in South Korea's national police (Bruce Cumings, *Korea's Place in the Sun: A Modern History* (New York: W.W. Norton 1997), 201).

Yet Rhee did not totally ignore the "mnemonic urban landscape" of his capital city. In a similar vein with acting to fulfil his ambition for personal power by emasculating the political system and turning it largely to his own,[52] when Rhee chose to employ colonial memories he did it out of personal considerations and through personal manifestations. In 1959, for example, the government came up with a plan to build a stadium for the upcoming 1960 Asian Cup football tournament. The site chosen was Seoul's Hyoch'ang Park where several independence fighters are buried, including Kim Ku who had become Rhee's most outspoken political rival at the aftermath of the liberation. On 26 June 1949 Kim was assassinated, perhaps at the directive of the president himself. The plan to build the sports stadium included the relocation of Kim's grave to the suburbs, diminishing its potential to evoke, through its spatial centrality, memories and myths about Kim that might in turn be additional catalysts for an already existing dissatisfaction with the president and his regime.[53] The stadium was finally built, yet public objection influenced the relocation plan to be dropped,[54] and Kim's grave exists at Hyoch'ang Park to date.

Furthermore, Rhee had two statues of himself erected at meaningful sites. One statue was placed in T'apgol Park (or Pagoda Park) at the heart of the city. T'apgol Park is where the anti-Japanese March First Independence Movement of 1919 started, and in South Korea this is the most revered manifestation of the anti-colonial struggle.[55] A second statue of Rhee stood on the grounds of Namsan ("south mountain") Park, which overlooks the city. The symbolism of this place goes back to the colonial period when Koreans were forced to worship at Japanese Shinto shrines, and among the hundreds of shrines that were peppered all over the country the biggest was placed at this dominating location.[56] Rhee's statue here towered 81 *ch'ŏk* in height (approximately 24.5 metres), symbolizing the

---

52   Eckert et al., *Korea Old and New*, 347–352.

53   For many years after his death, Kim's positions against division, foreign trusteeship, separate elections and US policy in Korea between the years 1945–1948, continued to be a problem for South Korea's governments that sought the friendship of the United States. These governments were also reluctant to advance knowledge about such positions because North Korea, too, held them at the time.

54   Kim Koo Museum and Library Website, www.kimkoomuseum.org/m02exhibition/ sub.asp?pagecode=m02s06 (in Korean) (accessed 28 May 2008).

55   Anti-Japanese demonstrations quickly spread throughout Korea in early March 1919 and caught the Japanese by surprise. Colonial authorities then resorted to brute force and extreme measures to quell this mass display of resistance at the expense of many thousands of Korean casualties, including over 7,500 dead. A good introduction to the March First Movement is Chong-sik Lee, *The Politics of Korean Nationalism* (Berkeley and Los Angeles: University of California Press, 1965 [1963]), 89–126. For an analysis of the commemoration of the movement and of T'apgol Park as a memorial site see Guy Podoler, "Revisiting the March First Movement: On the Commemorative Landscape and the Nexus between History and Memory", *The Review of Korean Studies* 8, no. 3 (2005): 137–154.

56   Cumings, *Korea's Place in the Sun*, 182.

president's 81st birthday, and it was known at the time as the tallest bronze statue in the world.[57] In the end, the fate of these two statues tells much about their meaning – both were brought down by angry protestors during the April Uprising that toppled Rhee's regime.

In conclusion, during the embryonic stages of South Korea, the country's development was hindered by the convergence of social, structural and institutional problems, with the North Korean threat and its tangible consequences, and with a president who, in the words of the historian Adrian Buzo, "had little sense of what an independent Korea might look like".[58] With Rhee's general lack of appreciation to the significance of a systemized national history,[59] and with the colonial stain on his administration, the president did not invest much in employing colonial memories for his service. In the few cases when he did so, he tried to create a personally based myth by placing his statue at a spot signifying the fight for independence,[60] while attempting to convey his authority by replacing a former colonial structure of domination with a postcolonial one at Namsan Park. Not only, then, was there no serious attempt in Seoul to use colonial memories as myths that could then have been passed down to future generations; but, also, the destruction of Rhee's statues signified that, at least for the time being, not much was to be expected with regard to a "Rhee myth".

The Park Chung-hee Era, 1961–79

On 16 May 1961 General Park Chung-hee took control over South Korea through a military coup, which put an end to the ill-fated attempt at democracy of the Second Republic (1960–61). In the two years that followed he ruled through martial law and laid the foundations for the recovery and development of his country. Then, under the constitutions of the Third and the Fourth Republics (1963–72 and 1972–79 respectively) Park initiated consecutive Five-Year Economic Development Plans and boosted industrialization. His unprecedented authoritarian rule in the 1970s was conducted under a constitution that "transformed the presidency into a legal dictatorship",[61] hence Park's controversial image of being either a great leader or a ruthless oppressor, or both. In any case, Park was a pragmatist who believed in, and carried out, institutional and structural systemization required for establishing a strong nation, and this included the systemization of writing history.

---

57   Chae-chŏng Chŏng, In-ho Yŏm, and Kyu-sik Chang, *Sŏul kŭnhyŏndaesa kihaeng* [Korean] (Seoul: Sŏulhak yŏn'guso, 1996), 231.

58   Buzo, *The Making of Modern Korea*, 106.

59   Yonung Kwon, "Korean Historiography in the 20th Century: A Configuration of Paradigms", *Korea Journal* 40, no. 1 (2000): 45–46.

60   Rhee himself did not participate in the March First Movement. During the colonial period he was outside of Korea and his activities for the sake of independence were political and diplomatic efforts.

61   Eckert et al., *Korea Old and New*, 365.

Under Park, national history was established through the education system, the works of "nationalist" historians and the erection of museums and memorial halls. It would be safe to assume that Park truly believed that becoming a strong nation entailed the consolidation of the sense of national identity through a consensual perception of the past. Yet by heavily investing in the promotion of a national history, Park also intended to muster support for his regime's claim for legitimacy as well as for its economic strategies.[62] Accordingly, Park's governing mythology focused on the history of Silla – the ancient kingdom that unified the Korean peninsula in 668 CE – because, first, Silla defeated the northern kingdom of Koguryŏ – the kingdom that North Korea claims for its own legitimization – and, second, Silla had originated in the same region that was Park's home and power base.[63]

Under this context, Park also resorted to the figure of Admiral Yi Sun-sin, the war hero who had bravely fought the Japanese invasions of the late sixteenth century. The president portrayed himself "as a late-twentieth-century Admiral Yi, one who saved the nation from Communist threat and an unfavorable international situation".[64] Thus, in 1968 Park erected what has become a Seoul landmark: a 19-metre-high statue of the admiral, dominating the Sejong-no–Chong-no intersection, which is one of Seoul's busiest traffic arteries. In February 2004, as the online edition of *The Korea Times* from 16 February reported, Seoul City officials announced that Yi's statue would be moved to a nearby park by the spring of 2005 due to the facelift that was scheduled to take place in Sejong-no. At the centre of the public debate that ensued was not the statue as signifying the memory and myth of Park Chung-hee, but rather, the monument as part of the myth of the revered admiral. In the end, the relocation plan was shelved.

With regard to colonial mythology, it should be noted at the outset that, like his predecessor, Park too was burdened by a difficult collaborationist colonial past: in the early 1940s he was admitted to, and served in, the Japanese army. Yet in addition to advancing a governing mythology that was based on ancient history, Park also systematically treated the colonial past. For example, already in 1962 selected independence fighters from the colonial period began receiving from the government, some posthumously, the National Foundation Medal of Merit of the

---

62 Andre Schmid, *Korea Between Empires, 1895–1919* (New York: Columbia University Press, 2002), 267.

63 Ibid., 272. The city that was most affected by Park's Silla-oriented governing mythology was Kyŏngju, the former capital of Silla located at the southeastern part of the country. Large-scale excavations took place there during Park's time, unfolding the history of the ancient kingdom by the discovery of a myriad of relics and artifacts. New museum facilities were constructed, and the city's branch of the National Museum was elevated to Kyŏngju National Museum in 1975.

64 Gi-Wook Shin, "Nation, History, and Politics: South Korea", in *Nationalism and the Construction of Korean Identity*, ed. Hyung Il Pai and Timothy R. Tangherlini (Berkeley: Institute of East Asian Studies, University of California, 1998), 154.

Republic of Korea. And with regard to the "mnemonic landscape" of the capital city, a shift was evident.

To start with, in 1966–7 two of T'apgol Park's most distinctive features existing to date were established. The first is a statue of Son Pyŏng-hŭi who was one of the initiators of the March First Movement. His statue, which dominates the entrance, was placed on the empty pedestal that had once supported President Rhee's statue. The second element was 10 2-meter-high bronze bas-reliefs that depict scenes from the March First Movement. Also, another site that was invested in was Namsan Park, where a series of statues was erected on its grounds, creating a combination of figures from the colonial period with figures from other historical periods. Finally, a set of memorial sites valorizing figures from the colonial period appeared since the early 1970s. These included: Patriot An Chung-gŭn Memorial Hall in Namsan Park, where at the entrance there is a stone that bears the carved handwriting of President Park, reading "a shrine/sanctuary of true national spirit"; Tosan Park to commemorate An Ch'ang-ho; the restoration of a house associated with the life of Yun Pong-gil; and the Yu Kwan-sun Memorial Hall in Ewha Girls High School.[65]

To conclude, during the time of Park Chung-hee the attention given to the colonial period signified a shift from earlier trends. Notwithstanding the personal "Japanese past" of the president and the state's "collaborationist record" – both which were at the time stumbling blocks for the full blooming of colonial mythology – governing mythology did allocate space for colonial history. This trend grew parallel to Park's shift to a tougher authoritarian rule in the late 1960s and the early 1970s. Then, domestic and international pressures had mounted to create a sense of insecurity for Park,[66] and against this backdrop he appropriated, at least to a certain degree, the colonial past for his service. This act intended to induce cohesiveness at a time of growing crisis, and it was carefully controlled so the president's personal troubled colonial past would not surface.

The Post-Park Era

From an historical perspective, the post-Park era can be divided roughly into two parts: before and after the transition to democracy in 1987/8. Yet despite the obvious differences between these two periods, a new trend pertinent to the construction of mythology spans them both.

On 26 October 1979 President Park was assassinated by the head of the Korean Central Intelligence Agency, and the person who took advantage of the new circumstances was General Chun Doo-hwan. Chun's impressive show of force ultimately brought him to the president's seat from where he headed a much-hated authoritarian regime until 1988. For our purposes, a significant event that occurred

---

65   For a detailed analysis of these memorial sites and the figures commemorated there see Guy Podoler, "Space and Identity: Myth and Imagery in the South Korean Patriotic Landscape", *Acta Koreana* 10, no. 1 (2007): 1–35.

66   See Eckert et al., *Korea Old and New*, 363–365.

in June 1982 should be mentioned: the "textbook controversy" with Japan. The controversy broke with reports that forthcoming school textbooks in Japan would whitewash Japan's colonial past through soft terminology. As the debate developed, anti-Japanese feelings in South Korea soared and were expressed by an angry and emotional mass media and fierce street demonstrations.

With respect to the reaction of the government, two significant factors should first be underscored: first, Chun's regime suffered from an acute legitimacy problem, mainly because of the killing of some 200 citizens during the anti-government demonstrations in the city of Kwnagju;[67] and second, the fact that Chun, as the historian Bruce Cumings pointed out at the time, was the first leader without a connection to the colonial period.[68] These factors were instrumental in driving the president to join in with the public and form a united front against the former colonizer. Thus, the government not only pursued a hard-line policy on the issue, but it also arrived at a decision that in retrospect has made a big impact on the commemorative landscape: the decision to build Independence Hall – a monumental memorial site for the colonial period located about 95 kilometres south of Seoul. The Hall, which opened in August 1987, is the biggest museum and memorial site in South Korea and one of the largest of its kind in the world.

Independence Hall would soon become the harbinger of a wave of more sites that are dedicated to the colonial past, sites where a common theme is the juxtaposition of Korean valour and suffering with Japanese brutality. In the capital city, the "mnemonic landscape" was shaped by the construction of related halls, monuments and statues among which the following projects are worth mentioning: Martyr Yun Pong-gil Memorial Hall (opened in December 1988); Ŭiyŏlsa, "shrine of heroism", in Hyoch'ang Park in 1989/90; Sŏdaemun Prison History Hall (1998) as a part of Sŏdaemun Independence Park, which was renovated from 1988 and throughout the early 1990s; and the Memorial Hall of Tosan An Ch'ang-ho in Tosan Park (completed in 1998).[69] Also, Hyoch'ang Park and T'apgol Park were officially designated as Historic Sites in 1989 and 1991, respectively. It must be noted in this regard that on some occasions the "upgrading" of the colonial past within the nationalist myth is employed to bolster the ideological and historical stance of South Korea in the face of its northern sister, while, conversely, at other times it is employed to promote the notion of a divided nation that bears a common history.

---

67   The Kwangju Uprising that erupted on 18 May 1980 is a defining moment in modern South Korean history. Army forces clashed with demonstrators for ten days until quelling the protest at a tragic expense. A good study in English on this event is Gi-Wook Shin and Kyun Moon Hwang, eds, *Contentious Kwangju: The May 18 Uprising in Korea's Past and Present* (Lanham, MD: Rowman and Littlefield, 2003).

68   Bruce Cumings, "The Legacy of Japanese Colonialism", in *The Japanese Colonial Empire, 1895–1945*, ed. Ramon H. Myers and Mark R. Peattie (Princeton: Princeton University Press, 1984), 479.

69   Details on these projects appear in Podoler, "Space and Identity".

Several other projects that took place need to be presented in the context of South Korea's transition to democratic government, which occurred on 29 June 1987. Then, influenced by, first, intense clashes in the streets, second, the concern that domestic instability would result in the cancellation of the Seoul Olympic Games, and, finally, American pressure,[70] the government announced an eight-point declaration that signalled the end of authoritarianism. After the transition, more voices were allowed to be heard and, specifically, more alternative views and more contention pertaining to the significance of memorial sites surfaced.

For instance, in the early 1990s a heated public debate erupted over the question of whether to destroy the former colonial Governor-General Building. Built between 1916 and 1926 the modern edifice served the Japanese governor of Korea, and it was deliberately situated in front of Chosŏn dynasty's main palace, Kyŏngbok, so it would obstruct its view. After liberation it continued to dominate the centre of Seoul as postcolonial governments used it, turning it even into the National Museum in 1986. When Kim Young-sam (president, 1993–8) promised to demolish the building during his presidential campaign, the debate began. For Kim, the demolition symbolized that he would lead the country to a new era, and those who supported the act agreed that finally the time had come to cleanse colonial memory by bringing down the building that symbolized humiliation and submission. In contrast, those who objected pointed to the unnecessarily high expenses involved in the demolition project, and said it was a waste to destroy such a functional building. They also emphasized that as long as the building was still standing, it constantly reminds the people of their ability to overcome the troubled past. Finally the decision came, and with the erasure of this dominant structure the landscape in the heart of Seoul was significantly altered.

Other important "mnemonic projects" in Seoul dating from after the transition are, first, a large memorial hall for Kim Ku that opened in October 2002 after a much smaller one existed since 1991, and, second, the War Memorial that opened in 1994. The significance of the first is that colonial mythology can now be more diverse as private associations construct memorials even for Kim who was once regarded by the authorities as a problematic figure (see 131, this chapter). The second site, the huge War Memorial, is not dedicated to the colonial period, but instead it celebrates the nation's history of military and war heroism. Interestingly, "a minor, although vocal, public outcry" was created as dissident intellectuals and students claimed that the museum "signaled the continuation of state authoritarian power through the forced celebration of a patriotic history imposed upon the public from above".[71] In any case, since the early 1980s governing mythology converged with other myths and with memories in relation with the colonial past. Through the addition of military heroism, a narrative of a dogged resistance has unfolded, yet more possibilities have been opened before alternative voices as well.

---

70   Eckert et al., *Korea Old and New*, 383.

71   Sheila Miyoshi Jager, *Narratives of Nation Building in Korea: A Genealogy of Patriotism* (Armonk: M.E. Sharpe, 2003), 120.

Conclusion

Based on this study's analysis of the dynamics pertinent to advancing governing mythology through memorial sites in Seoul, it may be concluded that the ability of South Korea's authoritarian governments to "impose memory" at the expense of the preservation and the development of private and collective memories and myths was limited. Provided with favourable conditions later in time, the latter flourished hence contributing to the complicated process of identity formation, while accentuating the contentious feature of memorial sites that operate as showcases of myths and progress in an environment of inviting recreational areas.

Furthermore, as Seoul is nested in multilevel configurations, the endeavour to cement identity for the nation and/or for the city through the "mnemonic landscape" and under changing socio-political conditions, was influenced by, first, the condition of a divided nation; second, the ruling elites' motivation to improve their image and strengthen their legitimacy; and, third, the contesting alternative voices that notwithstanding either their success or failure in achieving their goals, further define national or city identity simply by existing. In sum, what urban landscapes have to tell us about national identity is that beyond the particular historical context under which this identity was shaped, a more evasive identity is formed through the existence of particularities that colour the "sense of sameness" with a variety of shades.

References

Books and Articles

Ahn Kunhyuck and Yeong-Te Ohn. "Metropolitan Growth Management Policies in Seoul: A Critical Review". In *Urban Management in Seoul*, edited by Won-Yong Kwon and Kwang-Joong Kim, 49–72. Seoul: Seoul Development Institute, 2001.

Barthes, Roland. *Mythologies*, trans. Annette Lavers. New York: Hill and Wang, 1972.

Bell, Duncan S.A. "Mythscapes: Memory, Mythology, and National Identity". *British Journal of Sociology* 54, no. 1 (2003): 63–81.

Buzo, Adrian. *The Making of Modern Korea*. London and New York: Routledge, 2002.

Chŏng, Chae-chŏng, In-ho Yŏm and Kyu-sik Chang. *Sŏul kŭnhyŏndaesa kihaeng* (Korean). Seoul: Sŏulhak yŏn'guso, 1996.

Clifford, Mark L. *Troubled Tiger: Businessmen, Bureaucrats, and Generals in South Korea*, Revised Edition. New York: M.E. Sharpe, 1988.

Cumings, Bruce. "The Legacy of Japanese Colonialism". In *The Japanese Colonial Empire, 1895–1945*, edited by Ramon H. Myers and Mark R. Peattie, 478–496. Princeton: Princeton University Press, 1984.

Cumings, Bruce. *Korea's Place in the Sun: A Modern History*. New York: W.W. Norton, 1997.

Douglass, Mike. "Urbanization and Social Transformations in East Asia". In *Culture and the City in East Asia*, edited by Won Bae Kim, Mike Douglass, Sang-Chuel Choe and Kong Chong Ho, 41–65. Oxford: Clarendon Press, 1997.

Eckert, Carter J., Ki-baik Lee, Young Ick Lew, Michael Robinson and Edward E. Wagner. *Korea Old and New: A History*. Seoul: Ilchokak, 1990.

Gillis, John R. "Memory and Identity: The History of a Relationship". In *Commemorations: The Politics of National Identity*, edited by John R. Gillis, 3–24. Princeton: Princeton University Press, 1994.

Hae, Un Rii. "Jongmyo (Royal Shrine): Iconography of Korea". Paper prepared for the ICOMOS 14th General Assembly and Scientific Symposium: Place—Memory—Meaning: Preserving Intangible Values in Monuments and Sites, Victoria Falls, Zimbabwe, 27–31 October 2003. Available at www.international.icomos.org/victoriafalls2003/papers/A2-7-%20UnRii.pdf (accessed 27 May 2008).

Hall, Tim. *Urban Geography*, 3rd Edition. London and New York: Routledge, 2006.

Hill, Richard Child and Kuniko Fujita. "The Nested City: Introduction". *Urban Studies* 40, no. 2 (2003): 207–217.

Hill, Richard Child and June Woo Kim. "Global Cities and Developmental States: New York, Tokyo and Seoul". *Urban Studies* 37, no. 12 (2000): 2167–2195.

Jager, Sheila Miyoshi. *Narratives of Nation Building in Korea: A Genealogy of Patriotism*. Armonk: M.E. Sharpe, 2003.

Kim, Eun Mee and Jean S. Kang. "Seoul as a Global City with Ethnic Villages". *Korea Journal* 47, no. 4 (Winter 2007): 64–99.

Kim, Tong-Hyung. "Seoul Sees Rapid Growth as Global Metropolis". *Korea Times*, 15 August 2005.

Kim, Won Bae, Mike Douglass and Sang-Cheul Choe. "Introduction". In *Culture and the City in East Asia*, edited by Won Bae Kim, Mike Douglass, Sang-Chuel Choe and Kong Chong Ho, 1–14. Oxford: Clarendon Press, 1997.

Kim, Won Bae, Mike Douglass, Sang-Chuel Choe and Kong Chong Ho, eds. *Culture and the City in East Asia*. Oxford: Clarendon Press, 1997.

Kwon, Won-Yong and Kwang-Joong Kim. "Introduction". In *Urban Management in Seoul: Policy Issues and Responses*, edited by Won-Yong Kwon and Kwang-Joong Kim, 1–14. Seoul: Seoul Development Institute, 2001.

Kwon, Yonung. "Korean Historiography in the 20th Century: A Configuration of Paradigms". *Korea Journal* 40, no. 1 (2000): 33–53.

Leach, Neil. "Belonging: Towards a Theory of Identification with Space". In *Habitus: A Sense of Place*, 2nd Edition, edited by Jean Hillier and Emma Rooksby, 297–311. Aldershot: Ashgate, 2005.

Lee, Chong-sik. *The Politics of Korean Nationalism*. Berkeley and Los Angeles: University of California Press, 1965 (1963).

Lee, Dong Yeun. "Consuming Spaces in the Global Era: Distinctions between Consumer Spaces in Seoul". *Korea Journal* 44, no. 3 (Autumn 2004): 108–137.

Lee, Jin-woo. "Bribery Scandal Mars Restoration Project". *Korea Times*, 11 May 2005.

Lee, Ki-baik. *A New History of Korea*, trans. Edward W. Wagner with Edward J. Shultz. Cambridge, MA: Harvard University Press, 1984.

Lee, Mu-Yong. "The Landscape of Club Culture and Identity Politics: Focusing on the Club Culture in the Hongdae Area of Seoul". *Korea Journal* 44, no. 3 (Autumn 2004): 65–107.

Lee, Yong-Sook and Brenda S.A. Yeoh, eds. *Globalisation and the Politics of Forgetting*. London and New York: Routledge, 2006.

Olick, Jeffrey K. and Joyce Robbins. "Social Memory Studies: From 'Collective Memory' to the Historical Sociology of Mnemonic Practices". *Annual Review of Sociology* 24 (1998): 105–140.

Pai, Hyungmin. "Modernism, Development, and the Transformation of Seoul: A Study of the Development of Sae'oon Sang'ga and Yoido". In *Culture and the City in East Asia*, edited by Won Bae Kim, Mike Douglass, Sang-Chuel Choe and Kong Chong Ho, 104–124. Oxford: Clarendon Press, 1997.

Park, Soo-mee. "Controversy Continues Over Chosen Cheonggye Sculpture". *JoongAng Ilbo*, 17 February 2006.

Park, Soo-mee. "Local Architects Lash Out as Bulldozers Raze Dongdaemun". *JoongAng Ilbo*, 25 January 2008. Available at http://joongangdaily.joins.com/article/view.asp?aid=2885521 (accessed 27 May 2008).

Podoler, Guy. "Revisiting the March First Movement: On the Commemorative Landscape and the Nexus between History and Memory". *The Review of Korean Studies* 8, no. 3 (2005): 137–154.

Podoler, Guy. "Space and Identity: Myth and Imagery in the South Korean Patriotic Landscape". *Acta Koreana* 10, no. 1 (2007): 1–35.

Sassen, Saskia. *The Global City: New York, London, Tokyo*. Princeton: Princeton University Press, 1991.

Schmid, Andre. *Korea Between Empires, 1895–1919*. New York: Columbia University Press, 2002.

Seoul Development Institute. "Infrastructure". In *Changing Profile of Seoul: Major Statistics and Trends*. Seoul: Seoul Development Institute, 2005. www.sdi.re.kr/nfile/about_seoul/contents/profile_seoul_4.pdf (accessed 28 May 2008).

The Seoul Times. "Mayor Lee's Surprise Remark – 'Will Use City Budget for Anti Capital Move' – Lee Fights Back Against Ruling Uri Party Attack". N.d. (2004?). Available at http://theseoultimes.com/ST/?url=/ST/db/read.php?idx=1049 (accessed 27 May 2008).

Shin, Chang-Ho and Chang-Heum Byeon. "New Industrialization in Seoul: Industrial Restructuring and Strategic Responses". In *Urban Management in Seoul*, edited by Won-Yong Kwon and Kwang-Joong Kim, 125–146. Seoul: Seoul Development Institute, 2001.

Shin, Gi-Wook. "Nation, History, and Politics: South Korea". In *Nationalism and the Construction of Korean Identity*, edited by Hyung Il Pai and Timothy R. Tangherlini, 148–165. Berkeley: Institute of East Asian Studies, University of California, 1998.

Shin, Gi-Wook and Kyun Moon Hwang, eds. *Contentious Kwangju: The May 18 Uprising in Korea's Past and Present*. Lanham, MD: Rowman and Littlefield, 2003.

Wang, Chia-Huang. "Taipei as a Global City: A Theoretical and Empirical Examination". *Urban Studies* 40, no. 2 (2003): 309–334.

Yim Seok-Hoi. "Geographical Features of Social Polarization in Seoul, South Korea". In *Representing Local Places and Raising Voices from Below*, edited by Toshio Mizuuchi, 31–40. Osaka: Osaka City University, 2003. Available at www.lit.osaka-cu.ac.jp/geo/e-frombelow.htm (accessed 27 May 2008).

Websites

Kim Koo Museum and Library website, www.kimkoomuseum.org/m02exhibition/sub.asp?pagecode=m02s06 (in Korean) (accessed 28 May 2008).

Seoul Metropolitan Government website, http://english.seoul.go.kr/index.html (accessed 28 May 2008).

Chapter 8

"We Shouldn't Sell Our Country!": The Reconfiguration of Jewish Urban Property and Ethno-National Political Discourses and Projects in (Post)Socialist Romania

Damiana Gabriela Otoiu

"I arrived in Israel with a piece of paper saying: 'Free to pass. Citizenship – none' ... In order to be able to emigrate [in 1961], they made me sign a document by which I was giving up my Romanian citizenship and all the rights thereof",[1] an Israeli writer confessed to me, adding that she had no intention whatsoever to apply for Romanian citizenship or to claim ownership of assets which belonged to her family before the communist takeover. This is not an unusual situation: most of the 280,000 Romanian Jews[2] who went for *aliyah*[3] between 1945 and 1989 were forced to forfeit or to "sell" their properties to the Romanian state and some of them even to give up their citizenship. The same kind of "donations" were made by the Jews who emigrated to Europe or the United States.

Parallel to these so-called donations made by the members of the Jewish community, assets (especially real estate) belonging to Jewish institutions and organizations were appropriated by the state after the Second World War. Places of worship, schools, hospitals, retirement homes, soup kitchens, residential buildings, down to pianos, which could "be used in factories, for raising the cultural level of the new proletariat"[4] – a considerable number of goods belonging to Jewish

---

1 Interview with P. Transdnestr survivor, Israeli journalist and writer of Romanian origin, Tel Aviv, 30 December 2006.

2 These estimates were made by Radu Ioanid, *Rascumpararea evreilor. Istoria acordurilor secrete dintre Romania si Israel* [*The Ransom of Jews. The Story of the Extraordinary Secret Bargain between Romania and Israel*] (Polirom: Iasi, 2005), 203–204.

3 Word in Hebrew meaning the Jewish emigration to Israel (the literal meaning of the word is that of "ascension", "spiritual ascension").

4 Archives of the Federation of Jewish Communities from Romania (henceforth FJC Archives), Collection X, Movable and immovable assets, File 10, 1948–1950, f. 151 (the translations from Romanian or French into English are mine, unless otherwise mentioned).

communities were placed into state property and made available to the new political and administrative structures of the socialist state.

A central element of the strategy of the Romanian Communist Party,[5] which aimed to control and to bring an end to the life of religious communities, these nationalization acts were initiated as part of the "accomplishments of the regime of popular democracy that guarantees religious freedom".[6] For instance, the existence of educational institutions and of Jewish networks of social and medical assistance was said to be useless, considering that these areas of community life became a monopoly of the socialist state,[7] in the name of "the total equality between ethnic Romanians and all national minorities".[8]

After the breakdown of the socialist regime, the restitution of property, both individual and communitarian, became a key issue on the political agenda of Central and Eastern European countries. My chapter traces both these processes: of nationalization in socialist Romania (1945–89) and of (quasi)restitution of Jewish property in post-socialist Romania (as well as the polemics and the controversies which this restitution gave rise to).[9]

My analysis builds on two complementary approaches. On one hand, I'm looking at the expropriations that affected the Jewish community after the establishment of the socialist regime. In doing so, I'm trying to correlate the

---

5   As a general rule, to simplify, I use the generic name Romanian Communist Party (or, in short, Communist Party) to designate all the successive avatars of this political body: the Communist Party of Romania [Partidul Comunist din Romania], from its inception under illegality in 1921, the Romanian Workers' Party [Partidul Muncitoresc Roman], then created by the fusion of the Romanian Communist Party with the Social Democratic Party, in February 1948, and finally the Romanian Communist Party [Partidul Comunist Roman], after the name change in 1965.

6   Open Society Archives, Budapest (henceforth HU-OSA), Records 300 *Records of Radio Free Europe/Radio Liberty Research Institute (RFE/RL RI), 1949–1994*; Collection 60: *Romanian Unit, 1946–1995*; Series 1 (henceforth 300-60-1), Box 193, File "Jews 1957 – 1959", no. 1272/ 28.02.1952 : "Meeting of the Jewish Democratic Committee", in *Viața nouă* [*New Life*], 26.02.1952, 7.

7   Liviu Rotman, "The Politics of the Communist Regime Concerning the Jews: Contradictions, Ambivalence and Misunderstanding (1945–1953)", in *The Jews in the Romanian History. Papers from the International Symposium*, ed. Ion Stanciu (Bucharest: Silex, 1996), 231.

8   HU-OSA, Records 300-60-1, Box 193, File "Jews 1957–1959": document "Rumania. Freedom of religion acknowledged by Jewish official", 1.

9   My research is based on a fieldwork conducted in Bucharest and Cluj, Romania (2004–2008), in Tel Aviv and Jerusalem, Israel (2006, 2008) and on research in the archives (the archives of the Federation of Jewish Communities in Bucharest; National Romanian Archives, Bucharest; US Holocaust Memorial Museum Archives, Washington DC; the archives of the American Joint Distribution Committee, New York; Central Zionist Archives, Jerusalem; The Central Archives for the History of Jewish People, Hebrew University of Jerusalem and Open Society Archives, Budapest).

history of Jewish property with the history of Romanian "national Stalinism".[10]
I argue that the political and legal mechanisms of redefining Jewish property are
largely dependent on the history of the Communist Party. In this respect, one
needs to underscore that the history of the regime experienced at least two distinct
periods: a period of *Gleichschaltung*, of servile adoption of the Soviet model (end
of the 1940s, and beginning of the 1950s), followed by a period of simulated de-
Stalinization and de-Sovietization, of adopting a "national path" to communism
(especially starting in the 1960s).

On the other hand, I'm analysing the legislative framework concerning the
restitution of Jewish urban property after the fall of the socialist regime in 1989.
My analysis points to the fact that this process of re-privatization intersected
"ethno-national" political discourses and projects. The hypothesis is that the legal
redefinition of private property and the public debate around these laws could tell
us something about how the political elite "ethnified" restitution policies. Following
Offe,[11] I understand by ethnification of policies: "a number of interrelated strategies
of individuals and bodies, social as well as political. They are embedded in a cognitive
and evaluative frame according to which ethnic identity is a primordial and trans-
individual set of highly valued qualities that have been formed in a long collective
history". This implies that "policies are proposed, advocated and resisted ... in the
name of fostering the well-being of an ethnic community at the expense of excluding
those internal and external groups who are not considered to belong to it".

In the first part of this chapter, I will make an overview of the (apparently)
incongruous policies of the Communist Party concerning Jewish property between
1945 and 1989. In the second part of this article, I will focus on the reconstruction of
Jewish property in post-socialist Romania, by looking first at the legislative framework,
and subsequently at the political debates and controversies underlying the issuing of
these laws. I aim to give an answer to two interconnected questions: does the idea of
a homogeneous ethno-nation play a role in the elaboration of restitution policies? And
if so, who are the advocates of what Offe calls the ethnification of restitution policies:
exclusively the ultra-nationalistic parties or equally mainstream political actors?

**"Even if We Rule Them, We Don't Know What's Happening There ..." The
"Jewish Properties" in Socialist Romania**

"Damn them, they'll cry, but finally they'll shut up", said "comrade" Iosif
Chisinevski (a Jew himself) during a meeting of the Politburo of the central
committee of the Romanian Workers' Party from January 1953 that discussed

---

10 See especially Vladimir Tismaneanu, *Stalinism for All Seasons: A Political
History of Romanian Communism* (Berkeley and Los Angeles: University of California
Press, 2003).

11 Claus Offe, *Varieties of Transition. The East European and East German
Experience* (Cambridge, MA: The MIT Press, 1997), 51.

the possible abolition of Jewish communities and organizations. P. Borila, another member of the Politburo, continued: "we have the conditions to abolish these organizations of national minorities, which are an impediment for the development of our party's activity, for the mass education in the spirit of patriotism and proletarian internationalism ... The Jews and their organizations are very isolated from the party. Even if we rule them, we don't know what's happening there ... There is another problem, even more dangerous: the Jewish religious communities ... These communities have slaughter-houses for poultry, cemeteries, courses of catechization, Talmud ... We were thinking of letting the houses of prayer and the synagogues, but ... the slaughter-houses and the cemeteries must be taken immediately".[12] The leaders of the Romanian Communist Party, having reached power by *coup d'État* on 23 August 1944, with Moscow's blessing and support, were in the process of establishing a Stalinist regime in Romania. As they were adopting the Stalinist Soviet system, the nationalization of Jewish properties (and especially of the community assets, such as religious sites, schools, hospitals) was far from being an insignificant issue.

While the members of the Politburo of the Communist Party were secretly discussing the possible abolition of the Jewish organizations and the nationalization of their properties, they were publicly arguing for the necessity of a complete break with the pro-Nazi regime led by Antonescu, as well as with the "Romanianization"[13] process. Therefore, as in the case of other countries of Central and Eastern Europe, several laws were passed which were supposed to cancel the process of Romanianization and to proclaim the restitution of Jewish properties.[14] In fact, these laws were rarely put into practice, so most of the assets were not effectively restituted. The economic difficulties facing Romania right after the Second World War, and the fact that the "aryanized" goods had already gone into the possession of non-Jews, were obvious reasons for the limitations of the scope of restitution, sometimes even in spite of the existent legal framework. The difficulty of promoting potentially unpopular measures is even easier to explain when one keeps in mind the fact that in 1946 the politicians were preparing for elections.

One should mention, however, that a law on heirless and unclaimed Jewish property was passed in 1948.[15] This law allowed the Federation of Jewish

---

12   Romanian National Archives, Bucharest (henceforth RNA), Records of the Central Committee of the Romanian Communist Party (henceforth CC of the PCR), Section Chancellery, file no. 2/ 1953, "Minutes of the meeting of the Political Bureau, January 14, 1953", document included in Rotman, "The Politics of the Communist Regime", 238–242 (the translation of the document is made by L. Rotman).

13   The Romanian version of "aryanization" of properties (the seizure of Jewish properties) during the pro-Nazi regime of Antonescu.

14   For instance the Law for the abrogation of anti-Jewish legislative measures, adopted in Romania in December 1944.

15   Decree no. 113/ June 30, 1948, in *Colectiune de legi, decrete si deciziuni*, XXVI, 1–30 June 1948: 1527–1529.

Communities of Romania to retrieve all movable and immovable assets belonging to the Jews deported during the Second World War, to sell them and to use the money for "helping the poor". The "restitution" process was controlled by the Jewish Democratic Committee, an organization founded in 1945 by the communists in an effort to exercise better control over the Jewish community. The main task of this Democratic Committee was the "political enlightenment and mobilization ... of Mosaic clergy and the Mosaic faithful", which meant (according to the Committee's Secretary) "increasing the Jewish working people's love for and attachment to the U[nion of] S[ocialist] S[oviet] R[epublics], mobilizing Jewish working people more actively in the struggle for peace".[16]

Several local commissions were created in order to evaluate the situation of movable and immovable assets and to subsequently sell all the assets. The most important destination of this money, according to some confidential internal documents,[17] was the "restratification" of Jewish residents. The restratification was a social experiment by which the communists intended to re-professionalize the Jews engaged in commercial activities or in liberal professions, and to employ them afterwards in the industrial and agricultural enterprises owned by the state[18] (see Figure 8.1, a requalification centre in Bucharest, in the 1950s).

In the meantime, the country itself was going through an almost parallel process, (apparently) contradictory with the limited restitution of properties seized by the previous pro-Nazi regime: the nationalization of real estate, industrial enterprises, financial institutions, etc. The Jewish community was particularly affected by the laws directed at the teaching establishments (the nationalization of schools in 1948) and at the institutional networks for social and medical assistance (the nationalization of the hospitals, medical centres, asylums, canteens, in 1949). The education and the social services had to be the monopoly of the only existing party, as noted by Rotman.[19] Equally important, in 1948 the activity of the American Joint Distribution Committee was banned in Romania. Founded in 1914 as a humanitarian organization supporting several dozen non-American Jewish Communities, the Joint Committee had become by the mid 1940s the main finance provider for the Romanian Jewish Community, especially with regard to the Community's educational or social assistance programmes.

Similar to other Soviet satellite states, the fate of places of worship and belonging to various religious communities was a central topic of debate (and controversy) within the Romanian Communist Party political bureau. The synagogues were considered as being "the most dangerous" public space of the Jewish community:

---

16   HU-OSA, Records 300-60-1, File "Jews 1957–1959", no. 1272/ 28.02.1952, "Meeting of the Jewish Democratic Committee", in *Viata noua*, 26.02.1952, 2–7.

17   See FJC Archives, Collection X, Movable and immovable assets, File 11, "Minutes of the meeting of the Assets' Commission (Department of Cluj)", 30 August 1948, 18.

18   Tibori Zoltan Szabo, "Transylvanian Jewry during the Postwar Period, 1945–1948" (part 2), *RFE/RL Reports*, 6 (19), 2004.

19   Rotman, "The Politics of the Communist Regime", 231.

Figure 8.1 ORT building – Bucharest, circa 1950

Source: Photo Aurel Bauh, (ACSIER, VII 15, fm. 2), © CSIER-FCER, The Center for the Study of The History of Romanian Jewry – Federation of Jewish Communities from Romania, Bucharest.

> Jewish religious communities [are] the most dangerous. There are spies there … They have about 603 temples, houses of prayer, synagogues. The houses of prayer are capitalist enterprises for espionage … Our obligation of our state is to be firm, to take care of this problem: the synagogues must be reduced … As we have no pity for the catholic spies, let's not have any pity for the Rabbi.[20]

Romanian political leaders seemed to be torn between using the Soviet model of getting rid of places of worship and the model whereby places of worship were left undamaged, but kept under the watchful eye of the *Securitate* (the secret police, the Communist Party's main repressive institution). The individuals who replaced the former leaders of the Jewish communities and who were nominated or supported by the Romanian Communist Party were in favour of redefining the role of synagogues

---

20 "Minutes of the meeting of the Political Bureau of the Central Committee of the Romanian Working Party from January 14, 1953", reproduced in Rotman, "The Politics of the Communist Regime", 239–241.

in the new socio-political context. "After Soviet troops came [into Romania], the synagogue has to understand it has to go red", would have said one of the representatives of the Jewish Democratic Committee, rabbi Halewy, in a memorandum addressed to the Ministry of Religions.[21]

The solution of exercising political control over places of worship was ultimately preferred to their complete dismantling. The religious role of the Jewish community was in this way (apparently) preserved. However, the synagogues tend to be transformed into propaganda centres for the communist regime (for instance, strong anti-emigration and anti-Zionist propaganda).[22] "The *Securitate* [secret police] had plenty of informers present when religious service was organized in the synagogue … everybody in the community knew that it was common practice. On each High Holiday, some unknown people, usually men, came to the synagogue, people we knew weren't part of the community, and were there for one purpose only: to find out if we were talking against the system".[23]

After the gradual confiscation of most communal properties (e.g. schools, hospitals), and implicitly of some very important functions of the community (e.g. the education or the social assistance), pressure on communities decreased or it took on different forms than the violent repressive mechanisms of the late 1940s. Starting with the middle of the 1950s, the archives of the Federation of Jewish Communities show an ongoing process of negotiation between the leaders of the Federation, of the Communist Party, and the representatives of different institutions of the state (e.g. the Ministry of Religions, local administration).[24] The talks aimed, for instance, at re-establishing *yeshivot* (Talmudic schools), hospitals, community retirement homes or at allowing the activity of the Joint Distribution Committee, the most important finance provider for these activities before its banishment in 1948 (see Figure 8.2, the first visit in Bucharest of the representative of the Joint Distribution Committee, marking its reinstatement in Romania in 1967). Another issue that appeared on the agenda in the last decades of the regime was the fate of some of the major synagogues, which fell victim, like so many other historical and religious sites, to the "socialist systematization" of the country's cities. For instance, the federation managed to hold on to the Choral Temple in Bucharest,

21 Quoted by Slomo Leibovici-Lais, "Evreimea din Romania fata de un regim în schimbari: 1944–1950. Reinflorirea si lichidarea institutiilor evreiesti din Romania" ["The Romanian Jewry Confronted with a Changing Regime: 1944–1950. Recreating and Abolishing the Jewish Institutions in Romania"] (Ph.D. dissertation, Bar Ylan University, manuscript), 10.

22 Liviu Rotman, *Evreii din România în perioada comunistă. 1944–1965* [*The Romanian Jews during the Communist Regime: 1944–1965*] (Polirom: Iaşi, 2004), 39.

23 Centropa, Interview with Edita Adler (2003), www.centropa.org/archive.asp?mode=bio&DB=HIST&fn=Edita&ln=Adler&country=Romania, accessed June 2007.

24 See especially Central Archives for the History of Jewish People, Hebrew University of Jerusalem, Fonds: Romania - Moses Rosen and FJC Archives, Collection X, Movable and immovable assets and Collection VII M.R., Contemporary documents. Moses Rosen.

Figure 8.2   Charles Jordan in Bucharest, 1967

Source: Unknown photographer (ACSIER, VII 39, fm. 2), © CSIER-FCER, The Center for the Study of The History of Romanian Jewry – Federation of Jewish Communities from Romania, Bucharest.

a centre of religious life and the headquarters for the community, or the "Saint Union" Temple in Bucharest, which was turned, in 1977, into the Jewish History Museum. Most times, the price was the donation of places of worship or other community buildings, mainly somewhere else than in Bucharest.

The property rights of the Jewish community were dramatically curtailed by the 1980s as the "systematization" process was accelerated by Ceausescu. The communist regime tried to reshape cities by means of an aesthetic–political plan. For instance, Dudesti-Vacaresti (a Jewish area in the historic centre of Bucharest, the capital of Romania), was demolished almost completely, because of the architectural ambitions of the dictator Nicolae Ceausescu. He wanted to build a new "civic centre" dominated by the so-called "House of the People" (nowadays the Palace of Parliament, the second largest building in the world) and centred on the "Victory of Socialism" boulevard. More than half of the Jewish prayer houses in Bucharest were demolished between 1985 and 1988 during the building of the new civic centre.[25]

---

25   V. The chronological table of synagogues in Bucharest, published by architects Neculai-Ionescu Ghinea and Dan D. Ionescu in 1999 (consisting of some 110 temples/synagogues/places of worship built in Bucharest before 1985, of which 65 were demolished between 1985 and 1988), in *The Jews of Romania. History, Culture, Civilization* (CD

Jewish organizations were not the only ones forced to "donate" their assets, but individuals also. Despite the numerous hesitations in the Communist Party's politics on the question of emigration to Israel,[26] more than 280,000 Jews were allowed to make *aliyah* between 1948 and 1989. In the process they were forced to leave their possessions behind. So did the Jews wishing to emigrate to Europe or the United States: "Only those whose family pay a ransom of about 3–4 000 USD can leave [the country]. Obviously, they have to leave behind all the things they own, the house, the furniture ... In the case of the Jews who have no relatives abroad willing to pay the ransom, there is only one way ... they are supposed to make a donation to the State – the house and all their belongings".[27] Most probably, the authorization agreed to by communist leaders in relation to the problem of emigration was motivated economically: scamming the emigrants and their families, obtaining monetary compensations from the Israeli state and/ or international Jewish organizations (monetary advantages, "gifts" or "barters"), arranging economic schemes (help in obtaining international loans with zero interest or in concluding certain commercial agreements).[28]

The bureaucratic procedures for "ethnic" or "returning migration"[29] did not change dramatically during the socialist period. These migrations were only possible under two conditions. First, the Romanian leaders needed to negotiate in advance with high officials of "countries of origin" either directly or through mediators. The mediators were specialized institutions (such as the Jewish Agency, an organization founded by The World Zionist Organization, or the *Liaison Bureau*, an organization founded during the Cold War to encourage the *aliyah*, directly responsible to the Prime Minister of the Israeli State), non-specialized organizations (e.g. the humanitarian organization Joint Distribution Committee), diplomatic personnel, the secret services or simply persons that assumed this

---

edited by the Federation of Jewish Communities in Romania. The Centre for the History of Romanian Jews, Bucharest, 2004).

26   Rotman, "The Politics of the Communist Regime", 232.

27   HU-OSA, Collection 300 *Records of Radio Free Europe/Radio Liberty Research Institute (RFE/RL RI), 1949–1994*; Records 60: *Romanian Unit, 1946–1995*; Series 2 *RFE Confidential Reports on Romania* (henceforth Records 300-60-2), Box 3, Confidential reports Radio Free Europe (RFE), 1974: "Concerning the Situation of the Jews in Romania", 11 February 1974, 5.

28   For an extensive analysis of this process, based largely on unpublished sources, see Ioanid, *Rascumpararea evreilor*.

29   In this category, some theories on migration (see for instance Rogers Brubaker, "Migration of Ethnic Unmixing in the New Europe", *International Migration Review*, 32, 4 (1998): 1047–1065) include migrants perceived as returning to their "homeland". It is the case, as noticed by Michalon (Bénédicte Michalon, "Migrations internationales et recompositions territoriales en Roumanie: La propriété immobilière, enjeu des relations des migrants Saxons aux acteurs locaux en Transylvanie", *Méditerranée*, 3–4 (2004): 2), of the *Aussiedler* (people such as the Transylvanian Saxons, whose supposed "Germanity" gave them the benefit of the right to immigrate to the Federal Republic of Germany) or Jews who "ascended" to Eretz Israel (in Palestine, before the foundation of the State of Israel in 1948 and to Israel after 1948).

intermediary role (e.g. Henry Jacober).[30] These negotiations referred to the magnitude and timing of the immigration process, to the monetary compensations that the governments from the countries of origin (or other organizations) were supposed to pay to the Romanian state and sometimes to strategies of avoiding any publicity on the subject.

The second condition concerned the emigrant and consisted in some administrative steps that required the payment of several taxes, giving away real-estate goods (simply "donating" them to the state or selling them to a meagre price, lower than the actual price on the market). In certain cases future emigrants were even compelled to renounce their citizenship and/or the rights that were a consequence of that, such as the right to receive a pension.

Thus the economic consequences that sprung from the decision to emigrate were considerable. To the taxes for each administrative step (e.g. obtaining a passport, or a nationality identification card, giving up citizenship, obtaining the proof of not having any foreign accounts) and the "donations" or real-estate transactions on terms imposed by the Romanian state, quasi legal or illegal compensations to potential intermediaries and facilitators were sometimes added. Despite the fact that the "set price" for each emigrant-to-be varied between 826 and 10,000 dollars per person,[31] in reality the future emigrants were sometimes forced to offer much more. My respondents who emigrated to the United States at the beginning of the 1970s remembered:

> We finally managed to contact a specialist in these matters through my cousin who already emigrated to the U.S. ... The person let us know that our relatives already paid for our passports [15,000 dollars for my respondent and his spouse, n.n.]. However, there still is a small matter to conclude: "they" would like to buy the house. I said to myself, why would I need the house? ... and we left them behind, both in Bucharest and Falticeni.[32]

Property Restitution and (Ethno)National Projects

Like most Central and Eastern European countries, Romania instituted restitution policies after the collapse of the communist regime. However, the first restitution laws privileged the majority population, while excluding the non-citizens (emigrants who had lost or had given up their citizenship), the non-residents (citizens of a state who resided abroad), the former owners dispossessed before 1945 (it was the case for the Jewish properties "aryanized" during the Second World War), and the

---

30   For understanding the role played by Jacober in the Romanian *aliyah* (before and during his collaboration with the representatives of the *Liaison Bureau*), see especially Ioanid, *Rascumpararea evreilor* and Amos Ettinger, *Blind Jump: The Story of Shaike Dan* (New York: Herzl Press, 1992).
31   Ioanid, *Rascumpararea evreilor*, 164.
32   Email correspondence with D., New Jersey, July 2008.

"religious and ethnic minorities". For many political scientists, anthropologists or historians, this stemmed from the fact that, "in the post-communist world, restitution had become an adjudicator of national identity and ethnicity".[33] The legal status of property belonging to minorities was often part of "the agenda to keep all types of 'others' out".[34] Other laws, besides the restitution ones, particularly the constitutions passed after 1989, gave considerable advantages to the members of the "ethno-nation" – what Hayden called "constitutional nationalism".[35]

Property Restitution: Legal Framework ...

In Romania, the process of restitution of community property (assets which used to belong to different ethnic or religious communities and associations) nationalized during socialism was set in place after 1997, the date when the first normative acts regulating such matters were issued.[36] These first legal prescriptions were governmental emergency ordinances and allowed for the restitution of a limited number of communal and organizational Jewish assets.

When asked why the restitution was regulated for a long time by emergency ordinances, and not by regular laws, most of the members of the Parliament I interviewed gave me the answer: "the restitution of properties belonging to ethnic and religious organisations was not a very popular idea among the members of the Parliament, to say the least. A restitution law concerning these properties, although important back then for Romania on the path to NATO and EU accession, would have been rejected by Parliament ... or discussed for years and years".[37]

The first major law pertaining to the restitution of communal property was approved by Parliament in June 2002. The law no. 502/2002 concerned buildings that belonged to religious communities, nationalized between March 1945 (when the communist-dominated government came to power) and December 1989. This legal cut-off was considered by the representatives of the Jewish community prejudicial and exclusionary, because it left out the period between 1940 and 1945, when a considerable number of properties were seized.

As for the property belonging to individuals, the first (quasi)restitution law was passed in the mid 1990s, law no. 112/1995. It was strongly criticized, because it

---

    33   Elazar Barkan, *The Guilt of Nations. Restitution and Negotiating Historical Injustices* (New York and London: W.W. Norton and Co., 2000), 131.

    34   Barkan, *The Guilt of Nations*, 128.

    35   Robert Hayden, "Constitutional Nationalism in the Formerly Yugoslav Republics", *Slavic Review*, 51, 4 (1992).

    36   The restitution of Jewish communitarian assets is regulated especially through the Governmental Emergency Ordinances (GEO) no. 21/1997, modified by GEO no. 13/1998 and GEO no. 101/2000; GEO no. 112/1998; GEO no. 83/1999 and the Laws no. 501/2002, no. 66/2004.

    37   Interview with Z., member of the Judiciary Committee of the Romanian Parliament, Bucharest, 15 March 2008.

excluded from restitution an important number of properties. Without going into further details concerning different restitution laws that succeed one another starting with 1995, completing or contradicting each other, I wish to mention one relevant prerequisite. Romania, like most of Central and Eastern European countries, initially restricted restitution to current citizens. Because many of the former owners had to renounce their citizenship before emigrating, they became ineligible to recover their property unless they regained Romanian citizenship. Most of my interviewees complained about the relatively short period of time between the moment the law came into force and the deadline for claiming restitution, saying that they didn't have enough time to regain Romanian citizenship and to file a restitution claim.

A close examination of the legislative framework allows us to notice that the laws that have been drafted in the field of restitution, as well as the associated jurisprudence, lack consistency. This instability of the legal environment is due mainly to the fact that restitution laws often constituted the object of dispute and negotiations both among and within the various political parties.

... and Political Controversies

The rhetoric of the "people" and the "(ethno)nation" was used strongly during these political negotiations. While generally the members of the National Peasant Christian Democratic Party or of the National Liberal Party have tried to put restitution, including the properties belonging to religious and ethnic communities, on the political agenda, the most vigorous opponents were the members of the ultra-nationalist Greater Romania Party.[38] They argued that this retroactive, reparatory justice (giving back properties to their former owners) will in fact institute injustice, because of the displacement of the tenants currently living in nationalized houses. As a consequence, only a general redistribution of land and housing would really "deliver justice".

Their arguments were also built around the idea of "Romanianness": they claimed that the government should implement "ethnos"-based restitution policies in Romania. "We shouldn't sell our country! A country is a living organism. If we cut it into pieces, in concert with the foreigners, we will offend God ... We have already offended God" (said a member of the Greater Romania Party in June 1997). According to the political leaders of the ultra-nationalist parties, these "ethnos"-based restitution policies should ideally exclude the "Others" (e.g. different ethnic and religious communities, émigrés, foreign citizens).

The anti-Jewish diatribes of the leaders of the ultra-nationalist political parties with regard to restitution policies further explain the reasons of this (non-

---

38   Lavinia Stan, "The Roof over Our Head: Property Restitution in Romania", *Journal of Communist Studies and Transition Politics*, 22, 4 (2006): 1–26 and Damiana Otoiu, "Mémoire du communisme, acteurs du postcommunisme. Les associations des propriétaires et des locataires des immeubles nationalisés", *Studia Politica. Romanian Political Science Review*, IV, 4 (2004): 885–918.

restitution) imperative. The most popular anti-Semitic topos is the "conspiracy theory": the members of this party argue that Jewish lobby groups control the political decision-making in Romania, and influence the passing of restitution laws. "Important political leaders, able to make crucial decisions in the political and economical spheres ... are used by the Israeli secret service in order to have an impact on the decision-making process; they have succeeded in promoting Jewish interests against Romanian ones. The Jews' desire to acquire the world's capital has as a consequence the ambition to control the capital and the assets of each and every country", stated an anonymous journalist in a periodical edited by the Greater Romania Party.[39]

The conspiratorial myths are often associated with revisionist arguments. Thus, important leaders of the Greater Romania Party, such as historian Gheorghe Buzatu, are promoters of revisionist theories, while its affiliated publications have yielded a very consistent "database" of negationist statements.[40] By invoking Norman Finkelstein's arguments concerning the "industry of the Holocaust", the leader of Greater Romania Party, Vadim Tudor, explains that the Jewish community is exaggerating the extent of the Romanian Holocaust in order to get as many properties as possible:

> Someone is interested in depicting the Romanian people as a murderous people ... certain Jews are attempting ... to lay the blame the Romanian people ... Most times, the explanation is this: the next step after such accusation is to ask for restitution, be it collective or individual, minor or major, massive ... I published in my magazine, *Romania Mare [Greater Romania]* ... a document signed by representatives of some leagues from Israel that promote the restitution of goods and properties and by a few representatives of the Romanian Ministry of Foreign Affairs. The document asked for no less than 400,000 homes from Romania, maybe a house for every Jew that was supposedly killed in Romania, as expressed in a blasphemy, because that stone that was installed in front of the Coral Temple in Bucharest, saying that 400,000 Jews were assassinated in Romania, is a blasphemy, a gross misrepresentation of history ... I recommend a book ... "The Holocaust Industry"; it is written by an American Jew called Norman Filkenstein [*sic*]. It ... provides concrete examples of the very profitable character, in certain hands, of the gradual and artificial inflation of the figures of the Holocaust and of laying the blame on certain countries and peoples.[41]

---

39   "Romania became a Jewish colony" (author: "The Real Romanian Intelligence Service"), *Tricolorul*, 2008, no. 1280 (9 June). Most of the anti-Semitic articles published in the journals printed by the Greater Romania Party (*Romania Mare* and *Tricolorul*) are not signed or are attributed to fictitious organizations, such as "The Real Romanian Intelligence Service".

40   International Commission on the Study of Holocaust in Romania, *Final Report*, ed. Tuvia Friling, Radu Ioanid and Mihail E.Ionescu (Polirom: Iasi, 2005), 354.

41   Minutes of the session of the Senate, 2 April 2002, Political statement of Senator Vadim Tudor, www.cdep.ro/pls/steno/steno.stenograma?ids=5274&idm=2,06&idl=1, accessed July 2003.

A third major political myth that has been invoked by the representatives of extremist parties such as the Greater Romania Party against the restitution of Jewish properties is the myth of "Judeo-Bolshevism". As Tismaneanu notes,[42] for Romanian ethnocentric prophets, such as Vadim Tudor, communism has been a political regime imposed by the "foreigners", and Marxism-Leninism – an "alien" ideology, propagated by the same "foreigners": "It is in the name of these few thousand Jews that official requests are made for no less than 400,000 so-called Jewish properties in Romania … Where did they get these properties? How did they earn these assets? … let's be honest … Who brought Communism to Romania? Most of them were Jewish … they, these people, have plundered the entire national heritage … let no one exploit Romania's weakness … to cause a haemorrhage of the national heritage. Let no predatory ravens circle the wounded body of Romania" (said in an electoral speech in October 1996 by Vadim Tudor, then candidate to the presidency of Romania).[43]

But the ethno-national discourses are not the monopoly of the ultra-nationalist political parties. Other political leaders (namely the representatives of the Social Democrats, the most important left-wing political party) try to legitimize discriminatory restitution policies. "Is it worth despoiling those who are today living in poverty?", wondered Ion Iliescu, former Romanian president[44] during an interview given to a journalist from *Ha'aretz* in 2003. He concluded then that, due to the country's bad economic situation, "the Jewish property restoration requests should be either postponed or rejected". It is not only the prohibitive cost of restitution that explains, in the Social Democrat's view, the "necessity" of "postponing" the restitution of properties belonging to different ethnic and religious communities. Another element is what they considered "the possible dangerous consequences on inter-confessional and inter-ethnic relationships".[45]

Conclusions: From Political Discourses to Everyday Discourses

The policies concerning the restitution of private property after the fall of the socialist regime contribute not only to the redistribution of economic resources, but also to the (re)creation of a (political) economic community. Restitution laws

---

42   Vladimir Tismaneanu, *Fantasies of Salvation: Democracy, Nationalism, and Myth in Post-Communist Europe* (Princeton: Princeton University Press, 1998), 85.

43   Octavian Andronic, *Turneul candidatilor pentru functia de Presedinte al Romaniei, august-noiembrie 1996* (Bucharest: Alegro, 1996), also available in electronic form, www.editura-aleg.ro/index.php?module=pagesetter&func=viewpub&tid=1&pid=36, accessed September 2008.

44   Interview with Ion Iliescu, in *Ha'aretz*, 25 July 2003, www.haaretz.com, accessed July 2003.

45   Censure motion introduced by the members of the Social Democratic Party in the Romanian Parliament, 16 June 2005, www.cdep.ro, accessed June 2005. The Social Democrats made this statement when talking about the properties of the Greek-Catholic Church in Transylvania.

are, as remarked by Avineri, "not only a victory of norms of private property over communist ideology and practice, but also as a vehicle for the construction of post-communist national identity".[46]

Analysing the re-privatization of urban Jewish properties by means of legal texts or through political discourses, I argued that this process intersected with "ethno-national" arguments. Romanian citizens, whose civil rights, including the right to property, were re-defined after 1989, are often envisaged by lawmakers as "an organic nation, in an ethnic understanding of citizenship".[47] Thus, most of the restitution policies in Romania, similar to other post-socialist cases, are also the result of "ethnic politics", because they were created on the idea of a "special nexus among state, property, and national identity".[48] Moreover, this "ethnic politics" is largely consistent with norms and political practices inherited from the national-Stalinist regime.[49]

Exploring the laws and the explicit or implicit reasons underlying their elaboration, I tried to show that both nationalization and restitution processes have been intertwined with ethno-national discourses and projects. But certainly "aggregate statistics and compendia of decrees and laws tell us little without complementary close descriptions of how people ... are responding to the uncertainties they face".[50] Is this ethnification of restitution policies considered legitimate not only by some of the politicians, but also by their potential voters?

If one wants to see how popular "national identity" discourses are, the most accessible indicators are opinion pools and electoral results. Although Vadim Tudor's extremist Greater Romania Party ranks nowadays in opinion polls at less than 5%, below the required threshold for holding seats in the national Parliament, it's worth mentioning the fact that in 2000 he got one-fifth of the seats in Parliament using a stridently anti-Semitic, Hungarophobic, and anti-Western discourse. Or that in June 2009 Greater Romania Party, reunited with another party of the nationalist right, obtained more than 8% in the elections for the European Parliament, electing 3 MEPs. Or that Ion Iliescu, the political leader of the Social Democratic Party, convinced that "property is a trivial detail" and that

46 Shlomo Avineri, "A Forum on Restitution: Essays on the Efficiency and Justice of Returning Property to its Former Owners", *East European Constitutional Review*, 2, 3 (1993): 35.

47 Daniel Barbu, "De l'ignorance invincible dans la démocratie. Réflexions sur la transformation post-communiste", *Studia Politica. Romanian Political Science Review*, 1, 1 (2001): 23.

48 Katherine Verdery, "Transnationalism, Nationalism, Citizenship, and Property: Eastern Europe since 1989", *American Ethnologist*, 25, 2 (1998): 298.

49 For the analysis of the national(ist) ideology in socialist Romania, see especially Katherine Verdery, *National Ideology Under Socialism: Identity and Cultural Politics in Ceausescu's Romania* (Berkeley, Los Angeles and Oxford: University of California Press, 1995) and Tismaneanu, *Stalinism for All Seasons*.

50 Michael Burawoy and Katherine Verdery (eds), *Uncertain Transitions. Ethnographies of Change in the Post-Socialist World* (Lanham and Oxford: Rowman & Littlefield, 1999), 2.

the restitution of properties belonging to ethnic and religious groups might have dangerous consequences on "inter-confessional and inter-ethnic relationships", won three times in the presidential elections. But everyday discourses are even more relevant than opinion polls – how people describe the changing regimes of ownership, or how they narrate different individual experiences.

The judges, lawyers, politicians, former owners or tenants whom I interviewed[51] have different values and experiences about ownership. But many of the former owners (unsuccessful in getting back their properties) or tenants (risking a rent increase and eventually eviction) believe that the "foreigner" is "a materialistic descendant of industrialists and business owners who wants to become rich by exploiting our poverty" (R.V., 72). Or, as noticed by other informants, "they [the Jews] brought communism to Romania and then they escaped by emigrating to Israel ... Someone who ignored his origins all these years suddenly remembers and requests the return of their property, practically stealing it from Romanians ... How can this be called 'justice'?" (A.P., 57). Or "They [the Jews] always had a privileged status.[52] Maybe I should convert to their religion, so I can finally get my family's house back", says F.G. (45), adding that she gave her vote to Greater Romania Party both in 2000 and 2004 mainly because they promised to defend "ordinary people" from different "ethnic others" (such as Jews and Hungarians).

Obviously, these excerpts from my interviews do not (necessarily) reflect the way ethnicity and property are understood and constructed in all everyday practices and interactions. But sometimes, people who find themselves entangled in complex administrative and legal situations (confusions in property records, legal uncertainties, delays in the court system, etc.), looking for simple answers to complex problems, conveniently use the "foreigner" motif to explain these intricacies. Privatization is likely to be related to the theme of "(ethno)national identity" not just in political debates, but in everyday discourses as well. Besides, as Verdery[53] and others stressed, "part of what makes nationality so powerful is that it exists not just at the level of political rhetoric, interest groups and constitutionalism, but as a basic element of people's self-conception".

---

51   I made more than 120 interviews between 2004 and 2008, in Bucharest and Cluj (Romania).

52   She speaks about privileges granted to the "Jews" by referring to the laws pertaining to the restitution of religious properties. Despite the obvious distinction between communal and individual property, the confusion between these two types of property/laws is very common in the public discussions or in the daily discourse of my informants.

53   Katherine Verdery, "Nationalism and National Sentiment in Post-socialist Romania", *Slavic Review*, 52, 2 (1993): 194.

References

Andronic, Octavian. *Turneul candidatilor pentru functia de Presedinte al Romaniei, august-noiembrie 1996*. Bucharest: Alegro, 1996.

Avineri, Shlomo. "A Forum on Restitution: Essays on the Efficiency and Justice of Returning Property to its Former Owners". *East European Constitutional Review*, 2, 3 (1993): 30–40.

Barbu, Daniel. "De l'ignorance invincible dans la démocratie. Réflexions sur la transformation post-communiste". *Studia Politica. Romanian Political Science Review*, 1, 1 (2001): 19–28.

Barkan, Elazar. *The Guilt of Nations: Restitution and Negotiating Historical Injustices*. New York and London: W.W. Norton and Co., 2000.

Brubaker, Rogers. "Migration of Ethnic Unmixing in the New Europe". *International Migration Review*, 32, 4 (1998): 1047–1065.

Burawoy, Michael and Verdery, Katherine (eds). *Uncertain Transitions: Ethnographies of Change in the Post-Socialist World*. Lanham and Oxford: Rowman & Littlefield, 1999.

Ettinger, Amos. *Blind Jump: The Story of Shaike Dan*. New York: Herzl Press, 1992.

Hayden, Robert. "Constitutional Nationalism in the Formerly Yugoslav Republics". *Slavic Review*, 51, 4 (1992): 654–673.

International Commission on the Study of Holocaust in Romania. *Final Report*, edited by Tuvia Friling, Radu Ioanid and Mihail E.Ionescu. Polirom: Iasi, 2005.

Ioanid, Radu. *Rascumpararea evreilor: Istoria acordurilor secrete dintre Romania si Israel* [*The Ransom of Jews. The Story of the Extraordinary Secret Bargain between Romania and Israel*]. Polirom: Iasi, 2005.

Leibovici-Lais, Slomo. "Evreimea din Romania fata de un regim în schimbari: 1944–1950. Reinflorirea si lichidarea institutiilor evreiesti din Romania" ["The Romanian Jewry Confronted with a Changing Regime: 1944–1950. Recreating and Abolishing the Jewish Institutions in Romania"]. Ph.D. dissertation, Bar Ylan University, manuscript.

Michalon, Bénédicte. "Migrations internationales et recompositions territoriales en Roumanie: La propriété immobilière, enjeu des relations des migrants Saxons aux acteurs locaux en Transylvanie". *Méditerranée*, 3–4 (2004): 85–92.

Offe, Claus. *Varieties of Transition: The East European and East German Experience*. Cambridge, MA: The MIT Press, 1997.

Otoiu, Damiana. "Mémoire du communisme, acteurs du postcommunisme. Les associations des propriétaires et des locataires des immeubles nationalisés". *Studia Politica. Romanian Political Science Review*, IV, 4 (2004): 885–918.

Rotman, Liviu. "The Politics of the Communist Regime Concerning the Jews: Contradictions, Ambivalence and Misunderstanding (1945–1953)". In *The Jews in the Romanian History. Papers from the International Symposium*, edited by Ion Stanciu, 230–247. Institute of History N. Iorga, Romanian

Academy and The Goldstein – Goren Center, Diaspora Research Institute, Tel Aviv University. Bucharest: Silex, 1996.

Rotman, Liviu. *Evreii din România în perioada comunistă. 1944–1965* [*The Romanian Jews during the Communist Regime: 1944–1965*]. Polirom: Iaşi, 2004.

Stan, Lavinia. "The Roof over Our Head: Property Restitution in Romania". *Journal of Communist Studies and Transition Politics*, 22, 4 (2006): 1–26.

Szabo, Tibori Zoltan. "Transylvanian Jewry during the Postwar Period, 1945–1948" (part 2). *RFE/RL Reports*, 6 (19), 2004.

Tismaneanu, Vladimir. *Fantasies of Salvation: Democracy, Nationalism, and Myth in Post-Communist Europe*. Princeton: Princeton University Press, 1998.

Tismaneanu, Vladimir. *Stalinism for All Seasons: A Political History of Romanian Communism*. Berkeley and Los Angeles: University of California Press, 2003.

Verdery, Katherine. "Nationalism and National Sentiment in Post-Socialist Romania". *Slavic Review*, 52, 2 (1993): 179–203.

Verdery, Katherine. *National Ideology Under Socialism: Identity and Cultural Politics in Ceausescu's Romania*. Berkeley, Los Angeles and Oxford: University of California Press, 1995.

Verdery, Katherine. "Transnationalism, Nationalism, Citizenship, and Property: Eastern Europe since 1989". *American Ethnologist*, 25, 2 (1998): 291–306.

Chapter 9

Forgetting and Remembering: Frankfurt's *Altstadt* after the Second World War

Marianne Rodenstein

Figure 9.1   The *Altstadt* (old city) of Frankfurt, 1929
Source: Institut für Stadtgeschichte, Frankfurt am Main.

Introduction

In the days between 22–24 March 1944, the *Altstadt* (old city; see Figure 9.1) of Frankfurt am Main, Germany, was destroyed by an onslaught of bombing raids. In the autumn of 2007, Frankfurt's city council voted to reconstitute the core of the old city in accordance, more or less, to the old street grid of the Late Middle Ages and to rebuild around seven of the devastated residential and business houses from various periods of old-city construction using old plans, photographs and surviving fragments of the buildings, such as decorative elements. What led to this decision after some 63 years? Those structures of the old city seem to have been forgotten. Why, after so long, were they suddenly being remembered? What does city planning have to do with remembering and forgetting?

Forgetting and remembering are normal functions of memory, whereby without forgetting, which does not occur intentionally, as a rule, there can be no remembering. Both are bound up with the memory of the individual. In the case of Frankfurt's *Altstadt*, however, we are dealing not only with the memory of individuals but with that of the city itself, or rather, city policy and city planning. Whether a person forgets or remembers something clearly depends upon the significance of the event or thing for that individual. Applying this principle to the memory of the city, then, we might ask about the significance of the *Altstadt* for the self-image of Frankfurt before its destruction. This self-image expresses the power relations and future orientations of the city at particular junctures in its history. These are represented most visibly in its architectural structures. Relations of power change over time and are outlived, as a rule, by these structures. New power relations and the buildings erected within them produce new symbols. Old symbols are either forgotten or provided the occasion to be remembered. Which relations of power, which varying motives and mechanisms have influenced the forgetting and remembering of Frankfurt's *Altstadt* in the past 63 years?

In formulating an answer, we cannot disregard the social transformations in which cities are embedded, in which history and its architectural representations in the city functioned, now weaker, now more forcefully. In post Second World War (1939–45) West Germany, after the bombardment of Germany's cities, questions of the future and of reconstruction stood in the foreground. The historical old-city centres of many cities were completely wiped out. In only a few, such as Freiburg, Freudenstadt, Rothenburg and Münster, was there a consensus of opinion that the annihilated old-city quarters should be reconstructed – at least in part – in their previous form.[1] The political influence of monument conservators and traditionalist architects, as well as the interest of building owners in reconstruction, played a significant role in these cities. The process of forgetting and remembering is shaped by personal, local-political and professional, as well as social, forces. Each of the affected cities, thus, has its own individual history of old-city reconstruction, which begins with the question of the meaning of the devastated areas for the city itself and its own self-image.

The Meaning of the *Altstadt* for Frankfurt's Self-image

Frankfurt had been the site of the election and coronation of the Emperor of the Holy Roman Empire of the German Nation since 1356. The centre of festivities on the days of election and coronation was the city hall (or Römer, as it is called)

---

1   Werner Durth and Niels Gutschow, *Träume in Trümmern* (Braunschweig/Wiesbaden: Vieweg 1988), Volume I, 254.

and the cathedral not far from it. Frankfurt was under the immediate jurisdiction of the emperor, yet, as an imperial city, it enjoyed relative independence. The city was able to retain its independence even after the dissolution of the Holy Roman Empire of the German Nation and its transformation to the German Federation in 1815 by becoming a "free city" with statehood of its own.

In the Austro-Prussian war for dominance in Germany in 1866, however, Frankfurt was conquered by the Prussians and its status lowered to that of provincial town. As Wilfried Forstmann suggests, "In the history of the city, Frankfurt am Main, the year 1866 bears the weight of an epochal caesura. The violent end to Frankfurt's republican autonomy put a sudden close to an uninterrupted city-state tradition that had shaped the life of Frankfurt since the High Middle Ages".[2] Frankfurt lost its special, distinguished political status as a free city and as the assembling site of the sovereigns of the Federation, and had already lost its leading position in monetary trade. The very thing upon which the city's self-image had been founded was lost, or rather had been handed over to Berlin.

Industry now established a foothold in the city, which modernized itself in turn. Yet the continued presence of the late medieval *Altstadt* kept alive the memory of a past rich in tradition. While gothic structures were torn down to build a new street and, at the same time, expansions of the city hall were made in new gothic style the city purchased historically significant old-city structures, such as the Saalhof, the Palais Thurn and Taxis, the former seat of the Federal Assembly, as well as the Goldene Waage, an historic half-timbered house, in order to prevent the possibility "that these memorial sites from the city's distinguished past could be irreverently altered".[3] City policy-makers around 1900 were aware that the city needed to offer attractions to its visitors attending its trade fairs.

With its new self-image as a site of industry and trade, Frankfurt opened itself more than other German cities after the First World War (1914–18) to modernity in architecture and suburban housing. The city's new self-image in the 1920s was based on industry and architectural modernity, at the periphery, with modern residential structures for low-income social classes. Yet the city's history remained present in the form of the *Altstadt* at its centre. The tension arising from this juxtaposition of modernity and the Middle Ages characterized the atmosphere of the city. As a gathering site for the city's poor, and on account of its unhygienic living conditions and the genuine traffic obstruction it presented, it became clear, at this point, that the *Altstadt* with a population of 22,000 fit increasingly less into the self-image of a successful city of modernity and industry. Nevertheless, restoration was undertaken at several sites, which was continued by the city's National Socialist government. When the bombs fell on the nearly 2,000 buildings

---

2   Wilfried Forstmann, "Frankfurt am Main in Wilhelminischer Zeit 1866–1918", in *Frankfurt am Main. Die Geschichte der Stadt*, ed. Frankfurter Historische Kommission (Sigmaringen: Thorbecke 1991), 361. Own translation.

3   Friedrich Bothe, *Geschichte der Stadt Frankfurt am Main* (Frankfurt am Main: Diesterweg, 1913), 722. Own translation.

**Figure 9.2   The *Altstadt* with Römer (Town Hall) and Paulskirche (right),
1945**

Source: Institut für Stadtgeschichte, Frankfurt am Main.

of Frankfurt's *Altstadt* in 1944,[4] and costing over 5,000 lives, the image of the old
imperial city was destroyed (see Figure 9.2).

Forgetting and Remember Frankfurt's *Altstadt*

Examining the period between 1944 and 2008, there are five different phases of
public planning debates concerning the old city in which there are various occasions
and motives for forgetting and remembering which assume varying forms.

---

4   Werner Durth and Niels Gutschow, *Träume in Trümmern*, Volume II, 536.

The Restoration of St Paul's Church and Goethe's Birthplace

A few years after the destruction, no one had yet forgotten the old city; its image was still ingrained in the memories of its citizens. After the appointment of a new city administration by the American occupying forces in 1945, the city magistrate imposed a halt to construction of the fully demolished *Altstadt* so that nothing could be erected that might conflict with later design plans.

Initially, public discussion in the city was centred on the form of reconstruction of two highly symbolic old-city buildings, the church of St Paul's (the 100th anniversary of the first German democratic parliament in 1948) and the birthplace of Goethe (the 200th anniversary of his birth in 1949). How should these structures be rebuilt? Could the Germans' responsibility for the war and its consequences, of which the bombing of English cities as well as the destruction of Frankfurt's *Altstadt* were a part, simply be ignored and the buildings reconstructed in their previous form? Or, should the reconstruction take a form that documented the war, that kept its memory alive?

These are the questions that were publicly debated in Frankfurt in 1946/7 on the occasion of the reconstruction of St Paul's Church and Goethe's birthplace and which were ultimately answered by the city magistrate with a policy of "both/ and". The exterior of the reconstructed St Paul's Church – the site of assembly of the first German democratic parliament in 1848, until they were expelled by the troops of the reigning princes – strongly resembled the old church, while its interior reflected the spirit of contemporary times and the shortage of raw materials. The interior's spartan simplicity, however, was "not merely dictated by necessity, but was also a conscious decision to symbolize demonstrative humbleness at the beginning of a new historical period".[5] St Paul's Church was to become a symbol of Germany's spiritual rebirth.

At the same time, the group who had run the museum at the now demolished birthplace of Goethe was campaigning to rebuild the house exactly as it had been so that subsequent generations could see what a bourgeois home once look like.[6] And that is precisely what happened with the consent of the magistrate against the opinion of city planners. Those who opposed this historical reconstruction saw it as a denial of history. The destruction of Goethe's birthplace, so the argument goes, is part of the intellectual history of Germany and Europe and even this last chapter must be acknowledged. Moreover, even if it is difficult, the understanding that the demise of the house was a direct result of wrongdoing must be demonstrated rather than the defiant and sentimental production of the past, as Walter Dirks, editor of the *Frankfurter Hefte*, a critical intellectual magazine, argued in 1947.[7]

We see here already that the city's past served as a reservoir of justification for the desired form of reconstruction. While some wanted to rebuild in a way that

---

5　Ibid., 485. Own translation.

6　Ibid., 486.

7　Ibid., 488.

would keep the memory of the war and its wrongdoing, for which the Germans were responsible, alive, others wanted to use Goethe's house as a reminder of the humanity and humaneness that represent the better side of the German people. Both groups, however, wanted to create memorials with the respective structures that not only benignly commemorate the past but also serve to warn future generations. Different ways of dealing with Germany's National Socialist past and the war stood in the background concerning the question of reconstruction.

In the 1960s, when people began to wonder about how little National Socialism and the crimes associated with it were thematized in the postwar years, the explanations of the psychoanalyst Mitscherlich[8] became popular, according to which, for many, survival after the war was only possible through the denial and repression of their own participation in National Socialism. Accordingly, all blame was placed on the once-loved Führer, so that National Socialism could be bracketed off as a short period of misfortune in German history. The reconstruction of the Goethe house could be interpreted as the result of such a psychic splitting or repression, if, as with Mitscherlich, we want to apply individual psychological phenomenon to society as a whole.

Competition: 1950

After the division of Germany into two zones – the Russian occupied East zone and the West zone occupied by the French, English and American allies – and the prospect of the establishment of a German state, which the allies had been pushing since 1948, Frankfurt had justified hopes of becoming the new provisional capital of West Germany. Working in its favour were its geographical location in the centre of Germany and the fact that the city was Germany's political centre until 1866. Thus, city policy-makers wanted to retain the ruined old-city quarter to house the administrative buildings of the capital to come, a plan, however, that raised objections from a group supporting the reconstruction and repopulation of the old city with its previous residents and according to the old street network and lots. This "Association of Active Friends of the Old City" campaigned for a cautious reconstruction that would give builders and their architects free reign within the boundaries of building regulations and good taste.[9] After the new German parliament of West Germany voted on 11 November 1949, to make Bonn, rather than Frankfurt, the provisional capital, the old city lost its meaning not only as a symbol of the earlier "first city" but as a site for new capital buildings.

There were two plans for old-city reconstruction under consideration that would have roughly preserved the familiar image of the *Altstadt*. The plans included, in part, the block structures. But the city administration worked on the assumption that the historical old city had been destroyed and the former, mostly impoverished,

---

8 Alexander Mitscherlich, *Auf dem Weg zur vaterlosen Gesellschaft. Ideen zur Sozialpsychologie* (München: Beck, 1963).

9 Werner Durth and Niels Gutschow, *Träume in Trümmern*, 496.

building owners need not be taken into account in the matter of reconstruction.[10] They themselves had contracted a team of independent architects to draft a plan for the reconstruction of the old city. This team designed the buildings exclusively in row-house style, which reflected the struggle beginning in the 1920s between modern and traditionalist architects, the latter of whom advocated the block, though it was considered unhealthy for its lack of air circulation and of sunlight; the row-house style, on the other hand, was perceived as democratic because it provided everyone with the same right to healthy living conditions with "green" areas right outside the door.[11] This struggle preoccupied a large segment of the public. In early 1950 a competition to draft plans for the reconstruction of the old city centre took place. The "Friends of the Old City" lost their battle. Along with the churches, several historical buildings, such as the Saalhof chapel, the Leinwandhaus, the Steinerne Haus and the city hall (Römer) were to be rebuilt, but the old city as a residential quarter as it once had been would not be erected. Memories of these structures could no longer be borne in mind. The guidelines of the jury spell out that "[t]he arrangement of structures in plans for the newly reconstructed old city centre may not attempt at any point to replicate individual (medieval) buildings".[12] Regulations for the Römerberg specified that the three gables of the city hall were to be retained as an emblem of the city, while the remaining walls of the square should be rebuilt in simple, contemporary forms to avoid gestures of historicization.[13]

The competition brought no final results. In various sections of construction, apartments were built in row-house style or in large blocks that had nothing in common with the earlier *Altstadt*. For the empty space between the Römer and the cathedral, once filled by narrow alleyways and numerous residential quarters housing stores and businesses – the heart of the old city (see Figure 9.1) – there was no convincing solution to be found. "Respect for the history of this place demands caution in all decisions", it said in a memoir from 1953.[14] Frankfurt city policy-makers were not sure what should be done at this location. As a result, the area between the cathedral and Römer, so steeped in history, devolved into a parking lot.

To explain the triumph of modern, contemporary architecture in this competition for reconstructing the old city, we can point – along with the greater historical significance attributed by city policy-makers to some parts of the old city as elements of the city's self-image – to the fact that the row-house style and flat roof characteristic of modern architecture and city planning was already highly regarded in Frankfurt even before the period of National Socialism. Since this type

10   Ibid., 497.

11   For a more detailed discussion see: Marianne Rodenstein, *"Mehr Licht, mehr Luft". Gesundheitskonzepte im Städtebau seit 1750* (Frankfurt/New York: Campus, 1988).

12   Hans-Reiner Müller-Raemisch, *Frankfurt am Main. Stadtentwicklung und Planungsgeschichte seit 1945* (Frankfurt/New York: Campus, 1996), 56. Own translation.

13   Ibid., 61.

14   Ibid., 70.

Figure 9.3   The new *Altstadt*, healthy and democratic, in the 1950s

Source: H. Rücker, Frankfurt, www.aufbau-ffm.de/serie/Teil11/teil11.html; 9, 25 April 2008.

of architecture and planning was scorned during the Nazi era and not permitted to be built, the time had now come for these architects in contrast to their more traditionalist colleagues and monument conservators. They had, in any case, the moral upper hand in light of the Nazi ban on modern architecture. Moreover, it was their view that this architectural form fostered the new democracy by offering equal opportunities for housing to citizens (see Figure 9.3). Thus, these architects considered themselves future oriented, while their traditional colleagues sought to venerate memories of the old city's past.

In other areas of Frankfurt's inner city, construction had already begun to erect the first high-rise structures on the basis of nineteenth-century groundplans of the city. The city had been given special permission to build high-rises for reasons of economic development, so that already in 1953 the first plans for such a structure were included in city plans. The architects of modernity had made their presence felt in Frankfurt.

Competition: 1963

In 1963 a competition for planning the area around the cathedral and Römer took place. The area should provide housing, a building of cultural benefit for all, yet some 50 per cent of the area provided offices for the urban development department of the city. Funding for the project, however, was not secured. First prize went to an architect group that seven years later was awarded a contract to build the Technische Rathaus[15] in connection with an underground parking garage and a

---

15   The Technische Rathaus (technical city hall) houses the planning and construction department of the city.

subway tunnel. Audible public protest arose in response as a result of the enormity of the construction and the three towers which were to be positioned directly in front of the gothic cathedral and which had not been part of the plans submitted to the competition in 1963. These protests were headed up by the "Friends of Frankfurt" – successor to the "Association of Active Friends of the Old City" – which served as a representative for other influential local associations. The "Friends of Frankfurt" were opposed to the over-dimensional Technische Rathaus to which they preferred, "for the vitalization of the historically valuable site ... an architechtonically representative structure" as an artistic-cultural centre (flyer from January 1970).

There was no mention of the *Altstadt* or remembrance of it here at all. Neither did the association demand its reconstruction. As a result of the protests, the towers were ultimately permitted to be only as high as the cathedral's roof (see Figure 9.5). Hans-Reiner Müller-Raemisch, former director of the city planning office, saw as the occasion for the protests against the Technische Rathaus with its towers not only the aesthetic slap-in-the-face that this building represented in front of the gothic cathedral, but understood them also in connection with the protests surrounding the West End, an inner city residential area where the sanctioned construction of high-rise office buildings[16] was pushing out residents. Citizens defended themselves against the increasing arrogance of city policy-makers and planners, and in 1971 the head of the department of urban development was forced to step down.[17]

Competition: 1979/80

The new city image with its high-rises and the associated political struggles had a destabilizing effect on the political power held by the Social Democrats (SPD) since the end of the war. The city had a reputation as hideous and unattractive for the workforce. The construction planned in 1963 for the area between Römer and the cathedral was not pursued any further as a result of the protests against the Technische Rathaus. Taking his lead from public protests, the Social Democratic mayor at the time proposed the historical reconstruction of the east line of the Römerberg Square. New conflicts arose as a result, pitting community policy, which sought to appease the demands of the citizenry, against both the architects of modernity and monumental conservators, who were on the same side this time, perceiving any such reconstruction as a sham. The SPD had hoped to save the 1977 elections with the reconstruction campaign, but also genuinely believed it would make Frankfurt more attractive, beautifying the modern face of the city perceived by so many as hideous. They lost the elections, however, and the Christian Democratic Union (CDU) took up the idea.

---

16   Marianne Rodenstein, "Von der Hochhausseuche zur Skyline als Markenzeichen-die steile Karriere der Hochhäuser in Frankfurt am Main", in *Hochhäuser in Deutschland*, ed. Marianne Rodenstein (Stuttgart/Berlin/Köln: Kohlhammer, 2000), 33–46.

17   Hans-Reiner Müller-Raemisch, *Frankfurt am Main*, 70.

Figure 9.4   Results of the competition 1979/80: A front of half-timbered buildings on Römerberg Square with the post-modern Kulturschirn behind

Source: Georg Kumpfmüller, Frankfurt.

In 1979/80 another competition for planning the cathedral–Römer area took place, whose result was a compromise: the east line of the Römerberg was to be restored as half-timbered structures with exposed timber frameworks, which had been hidden beneath shingles until the destruction of 1944, while one building for cultural uses and exhibitions was to be designed in post-modern style. The jury spoke of this relationship as "dialectical". It also saw the historical east line, with modern behind it, as sufficiently estranged from what had gone before as to be considered a new structure, thus not a sort of historical plagiarism (see Figure 9.4).[18] Under the sign of post-modernity, the historical east line could be incorporated as a legitimate citation of the past. The politics of "both/and" ensured that there was something for every taste.

In the course of subway construction, the ruins of a Roman villa and the walls of a Carolingian palatinate were discovered and subsequently turned into an archaeological garden. The gateway for the first architectural remembering of the earlier old city after over 30 years was the public critique of the predominant contemporary planning and architecture in the cathedral-Römer area. That this critique was taken up by community policy-makers was, in the first instance, the

---

18   Ibid., 346/7.

Figure 9.5   Technisches Rathaus between Römer and cathedral, 2008

Source: Rolf Oeser, Frankfurt.

result of the negative image of the city that was hurting its economy. In this instance, popular taste, with the help of politics, won out against the professionals, the architects, the monument conservators and city planners. For citizens and tourists alike, the Römerberg – after nearly 40 years finally recognizable as a city square again – has proved to be a new point of attraction for the city (see Figure 9.4).

Competition: 2005

As a result of construction flaws, the Technische Rathaus already had to be repaired in the 1980s. The demolition of the building was deemed necessary in the 1990s as a result of further faults in construction and the search began for new uses for the location. City planners proposed the construction of a substantial hotel on the site, but there was no consensus among the political parties for this solution. So, in 2005 a competition took place for planning the site. The competition coincided with preparations for a major exhibit celebrating the 650th anniversary of the *Goldene Bulle*, a document dating back to 1356, declaring the city as the site of the election and coronation of the Emperor of the Holy Roman Empire of the German Nation. Thus, competition rules spelled out that plans had to take into account, above all, the former path of the Emperor from the Römer to the site of coronation in the cathedral, as well as allot space for residential structures, stores, restaurants, offices and a hotel. The archaeological garden could, but did not have to, be built over. City planners wanted to increase the economic value of this part of the *Altstadt*. The coronation path between the Römer and the cathedral had actually once been a somewhat wider winding alleyway, which would now be restored. In the years between 1356 and 1792, 17 emperors were elected and walked the path from the Römer to the cathedral (since 1562) to be crowned. City planners believed with this reconstitution to have adequately acknowledged the growing significance of history in general and of Frankfurt's history in particular.

Critiques from Frankfurt's citizens during the public exhibition of the competition results was fierce, yet at first sporadic. Conditions had already been placed on the winning sketches of a Frankfurt architect by the jury itself to downscale its size, to put a pitched roof in place of the flat one featured in the design and to take greater consideration of the historical city groundplans. Still, the design was attacked from all sides[19] and the debate took a fully unexpected, surprising new direction after an alternative to the modern plan was made public: the model of the historical *Alstadt* buildings that had stood at this location until March 1944.

The engineer, Dominik Mangelmann, had built a model of the buildings of Frankfurt's former *Altstadt* between the cathedral and Römer as the final year project of his studies, developed on the basis of historical documents. His first political support came from the Youth Union of Sachenhausen, the youth organization of the CDU. Mangelmann introduced the idea of rebuilding the half-timbered houses to the assembly of the city council. Most of the city's population was taken completely by surprise; after 60 years, it seemed this image of the old city lost in the war had been erased from the city's memory forever. Beginning in mid-October 2005, one event after another took place in Frankfurt centring on the issue. A collective process of remembering the old city was put in motion, above all in the pages of Frankfurt's three newspapers and in numerous public events, where now even the names of well-known individual buildings of the lost old city, such as the Goldene Waage, could once again be heard.

No one had anticipated that the issue of the old city would strike such a nerve among the public. Politicians seemed to lose control of the debate, in part because opinions within the parties were so markedly divergent. A citizens' initiative was formed to support the reconstruction of the old city, calling themselves "Pro Altstadt", while the Youth Union Sachsenhausen began circulating a petition for a referendum for the reconstruction of the half-timbered houses, exerting pressure on its own party. To maintain political control of the discussion, a special committee of the city council was formed to deal exclusively with question of the old city, as a forum in which the various citizens' interest groups were able to express their opinions as well. For, in the meantime, supporters of the plans for modern construction had come together to form an association of their own. Policy-makers understood that any decision in this matter would have to have

---

19 This sketch of the coronation path now showed a wide, straight aisle leading directly to the cathedral tower. It thus created a visual relationship that had never existed in earlier times.

the additional legitimacy of the citizens themselves. While the SPD was clearly prepared to let the citizens vote for one of the three alternatives for reconstruction under debate, the CDU and Green coalition did not want to relinquish control and organized a planning workshop open to the public in October 2006, whose results, however, would not be binding for the city council.

While it provided feedback to city planners, politicians were still stumped in the face of the conflict between modern construction and an authentic reconstruction, which continued, as before, to revolve around aesthetics. In the end, it all boiled down to matters of taste.

Memory as Justification of Taste

In the course of a few weeks, three different alternatives for construction at the former location of the Technische Rathaus were put on the table:

- a contemporary, modern structure, yet divided into small sections;
- the reconstruction of all of the 50 former houses; and
- a representational style mix with the reconstruction of only some of the more important earlier buildings. The rest would be new designs.

The groups supporting these various solutions tried to enlarge their numbers over time; it became clear, however, that those in favour of reconstruction attracted the most followers. But, if taste is not dictated by investors and their architects or by policy-makers, according to whose taste should the cityscape be designed? The taste of professionals or lay people, of architects or citizens, of those who prefer the purist aesthetic of modernity or those with "no taste", as it was said, who prefer the picturesque memory? All three groups employed various memories of the forgotten old city to legitimate and win support for their aesthetic opinion.

In the decision-making process concerning the planning of Frankfurt's *Altstadt*, memory appears as a means through which aesthetic opinion is freed of their subjectivity and thus be better legitimated. For memory has to do with values and with the future, while simple questions of taste, when they diverge from the professional standpoint of specialists, hardly have a chance to be taken seriously. Pierre Nora offers a sociological explanation for the increasing value of memory. He maintains that the burning need for history, for remembering, since the 1970s indicates the absence of a living memory in the community transmitted by living people.[20] Moreover, the dissolution of numerous forms of social connectedness has led, in many ways, to a loss of a collective memory, a collapse of a living memory, a memory of experiences. Contemporary mobility, as a rule, makes it hard to establish and hand down a collective memory of place. A memory of Frankfurt's old city based on experience exists today only among Frankfurt natives over 70

20   Pierre Nora, *Zwischen Geschichte und Gedächtnis* (Berlin: Wagenbach, 1990), 11.

years old. The memory of Frankfurt's *Altstadt* has long since retreated to museums and archives, where it survived, largely ignored, even forgotten. In the words of Aleida Assmann,[21] it has become a reservoir of memory that can be called up when needed. If there are now groups with reasons or motives to call up this memory, it is endowed with a vital relationship to the present. Assmann calls this kind of revivified memory, where the connection to the past is broken and reconstituted anew, functional memory, which is characterized by selectivity, the attribution of values and future orientation and which fosters identity. Competing versions of the memory of a place, as in the case of Frankfurt's old city, are produced; they are struggles for the authority to define the meaning of a place, with which claims to power are expressed in the form of city design. Roughly summarized, three groups with varying social backgrounds and values each promoted their own selective constructions of the historical old city and called them back into memory for the future design of the cityscape.

The group of professionals, the architects and planners, but some of the local media and political parties as well, belong to those who constructed a minimalist and, above all, non-picturesque functional memory of the old city in their design ideas. While these ideas integrated old street grids and lot sizes, the new was to be recognizably new, for the doctrine of modernity demands that every epoch design the future with its own means. To legitimate this standpoint, they point to discredited memories of the old city, the site of demonstrations of National Socialist power. This remark was made once at the beginning of the debates and never again. It was presumably understood that this argument no longer carried as much weight in the face of at least 600 years of old-city history and, now, nearly 60 years after the end of the war, as did 1948 in the debates surrounding the reconstruction of the St Paul's Church and Goethe's birthplace.

What is visible for this group is the task of construction and not the value of picturesque memories embodied by the half-timbered houses. This view of the old city is the most selective, that is, it leaves out the earlier image of the city, creating space by integrating only select features – among others, small-scale building and similar plot sizes – into a new image of a contemporary *Altstadt*.

The second group, those wanting to carry out their construction of memory as a thorough reconstruction of the old-city buildings, is made up of young and old citizens alike, members of the Youth Union of Sachsenhausen, as well as one of many citizens' initiatives supported in large part by women, who advocate this memory of the old city as a residential area, but also members of homeland and history associations. Their hopes for the future consist not only in the reconstruction of the approximately 50 half-timbered houses that once stood on the site of the Technische Rathaus but in recreating the spatial feel of the lost old city, the living place of ordinary people. This view of the old city remains the least selective. This group is fed up with the patchwork solutions of the cathedral–Römer area and is trying

---

21   Aleida Assmann, *Der lange Schatten der Vergangenheit. Erinnerungskultur und Geschichtspolitiks* (München: Beck, 2006), 54–58.

to create unity and harmony between the Römerberg, with its gothic facades and design, and the gothic cathedral by filling in the space with the historical *Altstadt*.

Their motives are manifold. Some identify Frankfurt as their home, but feel its historical meaning has not been adequately valued since the 1970s in the wake of modernization and globalization. Other, less educated members of this group form their aesthetic opinion on the basis of emotional and visual criteria. Still others no longer have faith in modern architecture in the old city and expect from city design an opportunity for emotional anchoring. They also point to the fact that the area to be reconstructed is ultimately a small part of the city.

Members of this group belong to the middle classes, rooted Frankfurt natives but also some who, precisely because of their mobility, appreciate the significance of historical city spaces and their capacity to foster identity.

The third group, whose plans seek to rebuild only select, representative buildings of the historical old city – leaving out in the process the reality of the deteriorated old city, living quarters for the poor in the nineteenth and twentieth centuries – prefer a mix of old and new. This was the solution supported by the majority of city council members, not only because it was politically amenable to some degree for all sides, but also because it best suited its ideas of decorating the city with magnificent, representative elements of its historical image. Old is now to be built anew, since after the war historical structures were sacrificed to the demands of the economy for new structures. Here, too, people spoke of giving the city a face. The functional memory that was created by this group serves to raise the historical value of the city and distinguish it from other cities in the context of global competition by taking up only the highlights of Frankfurt's history whose significance extend beyond the region.

In this spirit, the city council decided in the autumn of 2007 that in the cathedral – Römer area up to seven well-known, representative old-city buildings from various epochs, such as the Goldene Waage and some that bear the memory of certain people and emotional stories, like the house of "Aunt Melber", where Goethe ran in and out as a child, should be rebuilt. These will be mixed with contemporary structures.

What is remarkable about this, for the moment, last phase of the reconstruction debate is that the memory of the old city was provoked by a model of the image of the old city, which was seen at first by the younger generations as "simply beautiful" and for which they became advocates. Only then did the avalanche of collective remembering begin. This shows that even the skyline of high-rises, unique in Germany, did not quiet the longing for history and its structural representation; much more, precisely as a result of the modern and, in many cases, faceless architecture, this longing was reawakened.

Summary

For decades, Frankfurt's city planning has upheld a uniform cityscape that on the most sensitive sites again and again stood for the contemporary and economically beneficial solution and for the forgetting of the old city. Its decisions stirred up many a hefty public protest, which pushed policy-makers to "both/and" solutions, because a consensus between the aesthetic opinion of lay citizens and professionals was impossible to reach – hence the outcomes in 1948/9, 1979/80 and 2007. In spring 2010 the beginning of de- and reconstruction of the area between the cathedral and Römer began.

References

Assmann, Aleida. *Der lange Schatten der Vergangenheit. Erinnerungskultur und Geschichtspolitik.* München: Beck, 2006.

Bothe, Friedrich. *Die Geschichte der Stadt Frankfurt am Main.* Frankfurt am Main: Diesterweg, 1913.

Durth, Werner and Gutschow, Niels. *Träume in Trümmern. Planungen zum Wiederaufbau zerstörter Städte im Westen Deutschlands 1940–1950.* 2 Vols, Braunschweig/Wiesbaden: Vieweg, 1988.

Forstmann, Wilfried. "Frankfurt am Main in Wilhelminischer Zeit 1866–1918". In *Frankfurt am Main. Die Geschichte der Stadt in neun Beiträgen,* edited by the Frankfurter Historische Kommission, 349–422. Sigmaringen: Thorbecke, 1991.

Mitscherlich, Alexander. *Auf dem Weg zur vaterlosen Gesellschaft. Ideen zur Sozialpsychologie.* München: Piper, 1963.

Müller-Raemisch, Hans-Reiner. *Frankfurt am Main. Stadtentwicklung und Planungsgeschichte seit 1945.* Frankfurt/New York: Campus, 1996.

Nora, Pierre. *Zwischen Geschichte und Gedächtnis.* Berlin: Wagenbach, 1990.

Rodenstein, Marianne. *"Mehr Licht, mehr Luft!" Gesundheitskonzepte im Städtebau seit 1750.* Frankfurt/New York: Campus, 1988.

Rodenstein, Marianne. "Von der Hochhausseuche zur Skyline als Markenzeichen – die steile Karriere der Hochhäuser in Frankfurt am Main". In *Hochhäuser in Deutschland. Zukunft oder Ruin der Städte?,* edited by Marianne Rodenstein, 15–70. Stuttgart/Berlin/Köln: Kohlhammer, 2000.

Chapter 10

From "Patrimoine Partagé" to "Whose Heritage"? Critical Reflections on Colonial Built Heritage in the City of Lubumbashi, Democratic Republic of the Congo[1]

Johan Lagae

"Whose Heritage"?

On 2 February 2005, the eve of the day on which the Royal Museum of Central Africa in Tervuren was to open its major exhibition *Memory of Congo: The Colonial Era*, a curious event occurred in Kinshasa, capital city of the Democratic Republic of the Congo. The equestrian statue of King Leopold II that had been lying in a warehouse on the outskirts of the city for more than 30 years was re-installed in the city centre. It was mounted on the pedestal in front of the main railway station, where the statue of King Albert I had once stood.[2] The statue had originally been erected in 1928 and during the colonial era it powerfully embodied Belgian colonial power in what was once the capital city of the Belgian Congo.[3] In the context of Mobutu's *authenticité*-policy that aimed at erasing all traces of colonialism, the statue had been dismantled in the late 1960s. The then Minister of Culture of the RDC, Christophe Muzungu, presented the re-installment of the statue as part of a larger, future project intended to remind the Congolese of their history, including the colonial past, because "a people without history is a people without soul".[4]

---

1 This text is based on ongoing research regarding colonial architecture and urbanism in the Democratic Republic of the Congo and informed by a series of fieldtrips to Lubumbashi in the period 2000–2007. It forms a more concise and slightly modified version of a paper that was originally presented at the 1st symposium of the Ghent Africa Platform (GAP) in December 2007 and consequently published in *Afrika Focus*, 1 (2008): 11–30. I kindly thank the editors of *Afrika Focus* for their kind permission to republish this text.

2 The monument could not be reinstalled on its original location, which was on the square in front of the Palais de la Nation, as in 2001 a sumptuous memorial for the late president Laurent Désiré Kabila was erected on that spot.

3 This statue is an exact copy of the equestrian statue erected at the side of the Royal Palace in Brussels in 1926.

4 Christophe Muzungu as quoted in A.P., "Rise and Fall of a Brutal King", *The Times*, 4 February 2005.

The Congolese initiative and the opening of the *Memory of Congo* exhibition received international press coverage. Both events made it to the front page of several leading newspapers in Belgium, being presented as indicators of a shared moment of "rediscovered memory" in the former colony and mother country.[5] However, in Belgium as well as in Congo, the re-installment of the equestrian statue caused divergent reactions. While the replacing of the statue was applauded by some, others regarded the action as a tribute to a "*genocidaire*", thus expressing a critical stance on the Belgian colonial enterprise that has been widely popularized ever since the publication of Adam Hochschild's bestselling 1998 book *King Leopold II's Ghost: A Story of Greed, Terror and Heroism in Colonial Africa*.[6] After less then 24 hours, the equestrian statue was again removed, for a profound "cleaning operation" as described in the official press statement. It has not been put back into place since. The statue still stands today in the courtyard of the Institut des Musées Nationaux du Congo, ironically one of the most secured and least accessible locations in Kinshasa.[7]

The replacement of the equestrian statue of Leopold II clearly demonstrates the strong potential of physical remains of the colonial era to trigger and (re-)activate colonial memories. Colonial memories are, of course, as much about forgetting as they are about remembering. History informs such processes in complex ways, inducing practices that produce "grey memories" or, put differently, mixtures in varying degrees of "white" and "black" memories that cannot be reduced to a simple juxtaposition, just as colonial history cannot be written simply in "positive" or "negative" terms.[8] Hence, it will be argued here that in order to develop a meaningful approach to the colonial built legacy, it is important to focus not only on its tangible aspects (through operations of documentation, conservation or even restoration), but also to seriously engage with the complexities of the intangible aspects linked to its history and its embedded memories.

One of the key notions that has emerged in recent, institutional debates on the topic of colonial built legacy is "shared heritage". The scientific committee of the International Council on Monuments and Sites (ICOMOS), dealing with colonial built legacy, for instance, is operating under the label "Shared Built Heritage". While the official mission statement of the committee does not explicitly explain the term, discussion with several of its members indicates that it was chosen to

---

5   Colette Braeckman, "Congo : la mémoire retrouvée", *Le Soir*, 4 February 2005.

6   Hochschild's book, that makes strong claims on the atrocities and violence during the Leopoldian era, has led to huge controversies in Belgium. It produced a quite unproductive debate between believers and non-believers that resonated strongly in the press coverage of the *Memory of Congo* exhibit. For Hochschild's critical view on the exhibition, see "In the Heart of Darkness", *The New York Review of Books*, 6 October 2005.

7   In the course of 2004 the statue was transported from the Office of Public Transport in Kinshasa to the site of the Institut des Musées Nationaux.

8   "Passés coloniaux recomposés. Mémoires grises en Europe et en Afrique", *Politique Africaine*, 102 (2006). Theme issue edited by Christine Deslaurier and Aurélie Roger.

underline that every initiative regarding this built legacy should be rooted in a dialogue in which former "colonizers" and "colonized" can engage on a basis of equality.[9] Referring to the notion of "patrimoine partagé" in an introduction to the recent volume *Architecture coloniale et patrimoine. L'expérience française*, edited by the French Institut National du Patrimoine, Bernard Toulier takes a more outspoken position by arguing that the colonial built legacy no longer belongs to those who built it, but rather to those who inhabit it. In his opinion, the latter should be left the choice and responsibility of deciding what should be transmitted to future generations. But he also points to the paradox that is inherent in such a process. For how are former "colonized" to appropriate a culture that in essence is "foreign" to them?[10] His remarks offer a first complication to an all-too-easy usage of the notion "shared heritage". In her contribution to the same book, Mercedes Volait adds another critical note by reminding that any involvement of colonial heritage inevitably confronts the sensitive status of colonial history, a past which in Europe – and in France in particular, Volait adds – is still often masked rather than unveiled.[11]

The episode of the re-installment and subsequent removal of the equestrian statue of Leopold II forms a powerful indicator of the tension that exists between "shared heritage" and "dissonant history". But it also reminds us of the fact that "artefacts are not static embodiments of culture but are, rather, a medium through which identity, power and society are produced and reproduced".[12] Cultural heritage is indeed always a "social construct" to which multiple values are ascribed in dynamic processes of (re-)appropriation and negotiation. By suggesting that the values attached to the legacy do not necessarily have to be shared between the former "colonizers" and "colonized", let alone among members of these categories, the episode of the equestrian statue blurs any straightforward understanding of the notion of "shared heritage". Moreover, as one case study in this chapter will make clear, artefacts do not have to be objects of a shared interest to possess potential as "heritage". But perhaps most importantly, we should remain critical of the

---

9　The mission statement points out that the ICOMOS International Committee on Shared Built Heritage "supports public and private organizations world-wide in safeguarding, management and documentation of heritage and promotes and encourages its integration in today's social and economic life" (ICOMOS SBH leaflet, 2003). On Icomos in general, see www.icomos.org.

10　Bernard Toulier, "Introduction", *L'architecture coloniale. L'expérience française*, edited by Bernard Toulier and Marc Pabois (Paris: Institut National du Patrimoine, 2005), 23. I use the translation "a foreign culture" for the French "une culture exogène".

11　Mercedes Volait, "'Patrimoines partagés': un regard décentré et élargi sur l'architecture et la ville des XIXe et XXe siècles en Méditerranée", in *L'architecture coloniale. L'expérience française*, op. cit., 121–122.

12　*Values and Heritage Conservation: Research Report*, Erica Avrami, Randall Mason and Marta de la Torre (Los Angeles: The Getty Conservation Institute, 2000), 6. For current debates on heritage issues, see also *The Ashgate Research Companion to Heritage and Identity*, edited by Brian Graham and Peter Howard (Aldershot: Ashgate, 2008).

notion of "shared heritage" as its usage might lead to an obscuring of the power mechanisms and structures at work in heritage practices and policies in former colonized territories. To start and unravel some of the complexities involved in such enterprises, any reflection on the colonial (built) legacy should therefore, I suggest, return to the basic, but fundamental questions raised by Stuart Hall when he discussed heritage as a discursive practice in the context of 1990s multicultural Britain, namely *whose* heritage are we actually talking about? *Who* is it for? And *who* is concerned by it?[13]

Towards an Alternative Way of Documenting Lubumbashi's Built Heritage

Addressing such questions in relation to colonial built heritage demands an approach that goes beyond the traditional standards of documenting built legacy through formal description and physical assessment that often isolate buildings from their urban and historical contexts, be they social, economic, cultural and/or political. It demands alternative ways of documenting the architecture and urban form of the colonial era, ways that seek to establish meaningful but sometimes complex relationships between built fabric, history and memory.

The colonial built legacy of Lubumbashi, the mining city in Congo's province of Katanga formerly known as Elisabethville, offers a particularly interesting case to discuss such an alternative approach to colonial built legacy. The city's urban and architectural legacy has already been the object of some scholarly interest,[14] and has more recently also been presented as a noteworthy "heritage". In September 2005, the French Cultural Centre in Lubumbashi organized the *Premières journées du patrimoine en Afrique francophone* at the initiative of its former director Hubert Maheux. Consisting of an exhibition of photographs, a conference and a series of guided tours, the initiative aimed at creating a local awareness of the architectural quality of the cityscape, arguing that it should be taken into account in future urban planning policies.[15] As part of a continued effort the Centre also published a small architectural guide on the city in 2008.[16]

---

13   Stuart Hall, "Whose Heritage? Un-settling 'The Heritage', Re-imagining the Post-Nation", in *The Third Text Reader on Art, Culture and Theory*, edited by Rasheed Araeen and Sean Cubitt (London: Continuum, 2002), 72–84.

14   Aspects of the architecture and urban planning of Lubumbashi have been studied in Ph.D. dissertations by Bruno De Meulder Bruno (Catholic University of Leuven, 1994) and Johan Lagae (Ghent University, 2002), but a comprehensive study of the city from this perspective is yet to be written.

15   See http://lubumculture.site.voila.fr/expopatri.htm.

16   *République Démocratique du Congo. Lubumbashi. Capitale minière du Katanga 1910-2010. L'Architecture*, edited by Hubert Maheux, Marc Pabois, Serge SongaSonga and Johan Lagae (Lubumbashi: Espace Culturel Francophone de Lubumbashi, 2008).

Figure 10.1   Demolished 1920s building in the city centre of Lubumbashi
Source: Photograph: Serge Songa-songa, 2007 (coll. Johan Lagae).

The latter initiatives have been important in raising discussion on the importance of this legacy, at a time when the city of Lubumbashi was undergoing a tremendous evolution. Indeed, in the period immediately following the 2006 elections the city was witnessing an economic upheaval unprecedented in the last decades that lead to a frenzied building activity and an immense pressure on the existing real estate, especially in the city centre. As a result, the urban landscape started to change at a dramatic pace. By the end of 2006, the first historical building in the heart of the city was torn down to make way for a new structure, while several complexes in the meantime have been topped with extra storeys with little or no respect for the existing architecture (see Figure 10.1). If the global economic crisis is currently slowing down (building) business in Katanga considerably, one can still only speculate what the city will look like five years from now.

Notwithstanding the merits of initiatives emphasizing the interest and importance of Lubumbashi's built legacy, the introduction in this context of a heritage policy that focuses mainly on material aspects for defining the value of the built legacy, however, can and should be questioned. Longstanding European practices in the domain of conservation and restoration of built heritage more often than not are underscored by a conception that tends to define architecture in terms of "monumentality", "durability" and "history", while formal and spatial inventiveness as well as technical innovation still are key criteria used in the selection of buildings and sites worthy of attention. Applying such terms and criteria to evaluate the built legacy of the colonial era ignores the fact that during colonial times these notions constituted the basis of argument put forward by European architects that sub-Saharan Africa was an architectural "no man's land". Within Western discourse, the "traditional" African built forms could not be described, let alone valued, by such a frame of reference. It is significant in this respect that in many cases heritage initiatives in former colonized territories have tended to privilege the architecture of the former European neighbourhoods over the immense built production in the so-called "native towns" that is considered of less or no importance, thus reiterating, albeit often unconsciously, a colonialist discourse.

If documenting the built legacy of Lubumbashi seems the first obvious step in developing a heritage policy for the city, caution should thus be exercised when establishing the frame of reference used to define the selection criteria of what should be documented in the first place. Rather than defining the value of the built legacy mainly in art historical terms, I argue for a multi-disciplinary approach. My take as an architectural historian on the built heritage question in Lubumbashi in fact draws on insights from recent architectural historiography that since the early 1990s has been influenced by postcolonial studies, while it is also informed by the emerging interest in African urban spaces in the domain of the social sciences.[17] But it is equally indebted to the fascinating work on *Lushois* urban memory, by scholars like Johannes Fabian, Bogumil Jewsiwiecki and members of the locally operating group "Memoires de Lubumbashi" such as Donatien Dibwe and Gabriel Kalaba.[18] My approach then seeks to point out, first, how, why and by whom Lubumbashi's built legacy was produced and, second, how it has been re-appropriated over time by a wide variety of agents, some of them simply re-using it as available hardware while others have invested it with new meanings through divergent, sometimes even conflicting, operations.

Such a positioning of Lubumbashi's colonial built legacy as a critical interface between colonial history and postcolonial memory, I argue, can provide a meaningful alternative to the format of a classical architectural guide, and offer a more sound starting point for a debate on the city's built heritage. For what is needed is not only a documenting of architecture through factual descriptions and visual material, but also an approach that allows for re-situating

---

17   See for instance *African Urban Spaces in Historical Perspective*, edited by Steven J. Salm and Toyin Falola (Rochester: University of Rochester Press, 2005).

18   For an overview of the "memory work" in Lubumbashi, see Bogumil Jewsiewicki, "Travail de mémoire et représentations pour un vivre ensemble : expériences de Lubumbashi", in *Tout passe. Instantanés populaires et traces du passé à Lubumbashi*, edited by Danielle De Lame and Donatien Dibwe dia Mwembu (Paris, L'Harmattan, 2005), 27–40.

buildings in their changing urban and broader political-cultural contexts, while simultaneously linking them to the subsequent urban societies that occupied and experienced these spaces. Rather than focusing on buildings as isolated artefacts then, I argue for a reading that considers them as complex historical documents. Apart from formal description or physical assessment, issues of location – i.e. the relationships of a particular building to its surrounding urban fabric or other buildings in its vicinity – of patronage and of use over time then become crucial elements of analysis. By shifting the perspective from "shared heritage" to "whose heritage", such an operation also allows us to critically question the "binary" character of the notion of "shared heritage" that all too often remains defined in terms of homogenized categories of (former) "colonizers" and (former) "colonized".

In this chapter, I will develop this argument by focusing on two particular sites in Lubumbashi. Through the first site, the former theatre building, the notion of "shared heritage" will be questioned by pointing out divergences within the community of former "colonized" in the postcolonial context. The second site, the Jewish cemetery, highlights how the heterogeneity within Lubumbashi's former white colonial community is currently re-surfacing in the way the city's built legacy is re-appropriated. In both cases, the discussion draws on the concept of "lieu de mémoire" which the French historian Pierre Nora introduced in the mid 1980s, defining it as "any significant entity, whether material or non-material in nature, which by dint of human will or the work of time has become a symbolic element of the memorial heritage of any community".[19] Even though the application of Nora's concept in African contexts has been highly contested, especially within the milieu of French Africanists,[20] it remains useful for architectural historians. For, as Hélène Lipstadt has argued, it reminds us, first, of the importance of the spatiality of memory, and, second, of the need to address not only the tangible but also the intangible aspects of built form.[21] Moreover, the capacity of "lieux de mémoire" to take up new meanings over time – a capacity that for Nora was one of the main elements that made "lieux de mémoire" exciting – is particularly helpful to chart ruptures and continuities from colonial to postcolonial contexts, as the work on Algiers by Zeynep Çelik has so convincingly illustrated.[22]

19   Pierre Nora, "Preface to the English-language Edition", in *Realms of Memory: Rethinking the French Past*, edited by Pierre Nora (New York: Columbia University Press, 1996), xvii.

20   See for instance Henri Moniot, "Faire du Nora sous les tropiques?", in *Histoire d'Afrique. Les enjeux de mémoire*, edited by Jean-Pierre Chrétien and Jean-Louis Triaud (Paris: Karthala, 1999), 13–26.

21   Hélène Lipstadt, "Review of Pierre Nora (ed.), Realms of Memory: the Construction of the French Past", *Journal of the Society of Architectural Historians*, 2 (1999): 243–245.

22   Zeynep Çelik "Colonial/postcolonial Intersections. Lieux de mémoire in Algiers", *Third Text*, 49 (1999): 63–72.

Figure 10.2   Theatre building, Lubumbashi, arch. Yenga, 1953–56

Source: Photograph: Johan Lagae, 2005.

Lubumbashi's Theatre Building, a Cultural Legacy or a Postcolonial "lieu de mémoire"?

The former theatre of the city of Lubumbashi is one of many public buildings in the city that is well preserved and still functioning, even if its use has changed over time (see Figure 10.2). It was originally built as part of a larger cultural complex that also encompassed a music school and a museum. The theatre terminates the vista of a wide avenue, the use of symmetrical axiality in its volumetric composition serving to give it a monumental presence in this particular urban setting. Well equipped and in tune with up-to-date standards of its time, its concept was innovative. The stage not only opens on an inner theatre room with over 600 seats, but also allows an audience capacity of 2,000 people assembled in an open-air patio on the back side of the main volume.

The building was designed by Claude Strebelle, who shortly after receiving the commission founded the architectural office Yenga (Swahili for "to build").[23] The

---

23   See Johan Lagae, "Claude Strebelle"/"Yenga", in *Dictionnaire de l'architecture en Belgique de 1830 à nos jours*, edited by Anne Van Loo (Antwerp: Mercatorfonds, 2003), 536–537/604.

Figure 10.3   Original interior view of the theatre building
Source: Archives of Claude Strebelle, Liège (courtesy Claude Strebelle).

sculptural design approach, which is even more outspoken in the adjacent museum building, a project of Yenga dating from 1959–61, testifies to a clear ambition to define a new, contemporary architecture for Africa that could offer an alternative for the orthodox tropical modernism that swept through the continent during the 1950s.[24] The decorative programme of the theatre building was in tune with this architectural ambition. The interior decoration was executed by local, Congolese sculptors and painters, such as Mwenze Kibwanga, Bela and others who had been trained in the famous art studio of Pierre-Romain Desfossés and the Art Academy, directed by Laurent Moonens.

The building invites evaluation as a structure of significant cultural importance, given the status of its designer within Belgian architectural history and the fact that its architecture as well as inner decoration immediately received national and international critical acclaim (see Figure 10.3).[25] Notwithstanding the sound overall

---

24   For a survey of orthodox tropical modernism, see Udo Kulterrmann, *Neues Bauen in Afrika* (Tübingen: Wasmuth Verlag, 1963) in which the theatre briefly is mentioned (p. 23).

25   Sabine Cornelis, "Naissance d'un académisme", in *Anthologie de l'art africain du XXe siècle*, edited by N'Gone Fall and Jean Loup Pivin (Paris: Editions Revue Noire,

Figure 10.4   Mural frescoes in the stairwells of the theatre building, nowadays disappeared

Source: Archives of Claude Strebelle, Liège (courtesy Claude Strebelle).

maintenance of the building and the survival of most of the inner decoration, the colourful mural frescoes depicting animals that decorated the stairwells have now disappeared behind a coat of white paint (see Figure 10.4). From a heritage perspective, one might thus raise the question as to whether action should be taken to try and restore these frescoes.

In order to provide an answer to such question, it is important to re-situate the theatre building in its historical context, as both its form and decoration speak of an underlying colonialist agenda. Built with financial aid and ideological support of the Union Minière du Haut Katanga and the railway enterprise B.C.K, both driving forces of the city's economy and crucial pillars of the Belgian colonial project in the Congo, the initiative for Lubumbashi's theatre building testifies of a deliberate policy of using culture as a strategic tool of colonization. In his inauguration speech the Commissaire du District of the time stated unequivocally that "to colonize means to project into space civilization", a statement in which "civilization" stood

2001), 164–167. For a Congolese perspective, see Celestin Badi Banga Ne-Mwine, *Contribution à l'étude historique de l'art plastique zaïroise modern* (Kinshasa: Editions Malayika, 1977), 76–89.

for cultural expressions coming from the *métropole*.[26] While the choice of having contemporary Congolese artists decorate the interior seems to express a genuine interest in contemporary African culture, it should also be remembered that these artists worked in an artistic milieu that was itself pervaded by the ambivalences of the colonial era, rather than providing a completely free forum for Congolese artistic expression.[27] Moreover, the work of the Congolese artists was limited to the interior decoration, while the commission for the main sculpture of the front façade was given to a Belgian artist, Claude Charlier, who had arrived in the Congo in the early 1950s and would became professor at the local art academy in 1956. This difference between in- and outside recalls the hierarchy between "indigenous crafts" and "colonizers' architecture" characteristic of colonial representations at international exhibitions and thus seems in line with the master–servant discourse that underscored colonial cultural politics in Congo, as it did elsewhere.

When assessing the heritage value of the theatre building, we should take into consideration that the theatre building was first and foremost a cultural institution oriented towards a white, mainly Belgian audience, as this forces us to identify for *whom* this work could have the potential of being (re-)appropriated as a cultural heritage. Furthermore, it is important to investigate the ways in which the consecutive re-uses of the theatre building over time have engendered considerable shifts in meaning, not only of the complex itself but also of its urban setting. In fact, immediately after Congo's independence, the theatre building was "politicized", when on 11 July 1960 the Congolese politician Moïse Tshombe declared the independence of Katanga and installed the Katangese parliament in the theatre, perhaps as it was the only building in the city fit for such a purpose. Depicted on the money bills of the Banque Nationale du Katanga, the theatre, however, soon became the symbol of this political event. Even if Katanga's independence was short-lived, national unity being restored in 1963, this symbolic connotation of the theatre reappeared in the early 1970s in a painting from *Histoire du Zaire*, a series of canvases by the Congolese painter Tshibumba Kanda Matulu that form a striking expression of "memory work". By including the inscription "On July 11, 1961 [*sic*], Katanga becomes inde[pendent]" Tshibumba explicitly relates the building to this particular episode in Congo's postcolonial history.[28]

---

26 "A Elisabethville. Inauguration du theatre", *Jeune Afrique*, 24 (1956): 41–42 (my translation). The full quote, in French, reads: "Coloniser, c'est projeter dans l'espace la Civilisation. Ce theatre, désormais, sera un témoignange de la nôtre et apportera –lui aussi– un remède à la stagnation en multipliant des contacts avec la Métropole spirituelle don't nous avons le culte et la nostalgie."

27 This milieu edited the art journal *Jeune Afrique*, of which Claude Strebelle was artistic director. For a recent critical Congolese view on this art production, see Isidore Ndaywel è Nziem, *Histoire générale du Congo de l'héritage ancien à la République Démocratique* (Paris: Duculot, 1998), 490–492.

28 The depiction should read "le 11 juillet 1960...", but then the *Histoire du Zaïre* should not be seen as an accurate survey of Congo's history. Its importance lies, as Fabian argues, in its capacity of presenting a powerful narrative of how the past is remembered via

The theatre thus offers a fine example of a "lieu de mémoire" whose meaning has changed radically from colonial to postcolonial times, speaking differently to former "colonizers" and "colonized". But what makes the building particularly interesting for our discussion of the notion "shared heritage" is that it can be regarded first and foremost as a Katangese, instead of a Congolese, "lieu de mémoire". Rather than being the symbol of a collective national identity – which was in fact the role originally assigned by Nora to a "lieu de mémoire", as he developed the concept to rewrite French national history – the theatre speaks of an explicit moment in time when Congo's national unity was under serious threat. When Mobutu resumed power and re-installed national unity, he quite significantly renamed the theatre the "Bâtiment du 30 juin", after the date of Congo's national independence.[29] Given the fact that the theatre nowadays acts as the seat of the Assemblée Provinciale led by the charismatic figure Moïse Katumbi Chapwe, and that rumours of a revival of Katanga's claim for more autonomy are currently in the air, one can only wonder about the extent to which the current political situation will affect the building's significance and meaning in the years to come.

The Jewish Cemetery of Lubumbashi, a Virtual "Lieu de Mémoire" of a Formerly Cosmopolitan Colonial Society

The Jewish cemetery of Lubumbashi, a site quite remote from the city centre, being situated next to the industrial zone of the city, on the opposite side of the railway, allows for a further critical assessment of the notion "shared heritage" (see Figure 10.5). The cemetery is impressive in size and counts several large tombs that testify to the once important position of the Jewish community in Lubumbashi's urban society.[30] But it also forms a powerful reminder of the cosmopolitan nature of the city's colonial past, a characteristic that invites us to rethink our understanding of Lubumbashi as being just another Belgian colonial city.

---

the present. Johannes Fabian, *Remembering the Present: Painting and Popular Culture in Zaire* (Berkeley: University of California Press, 1996), 105.

29   Michel Lwamba Bilonda, *Histoire de l'onomastique d'avenues et de places publiques de la ville de Lubumbashi de 1910 à nos jours* (Lubumbashi: presses universitaires de Lubumbashi, 2001), 51–52 and 64. Even though he pays attention to Mobutu's effort to erase traces of the Katangese independence struggle, Lwamba Bilonda fails to critically point out the particularity of changing the name of the theatre building in that context.

30   On the history of the Jewish community in Congo, see the popularizing book by Moïse Rahmani, *Shalom Bwana. La saga des Juifs du Congo* (Paris: Romillat, 2002). For a more profound historical analysis of the Jewish community along the railroad from Cape town, with focus on Zambia and some information on Congo, see Hugh Macmillan and Frank Shapiro, *Zion in Africa: The Jews of Zambia* (London: I.B. Taurus, 1999).

Figure 10.5   Jewish cemetery, Lubumbashi

Source: Photograph: Johan Lagae, 2006.

At first sight, Lubumbashi's urban landscape does highlight the presence of the three main actors of the colonial enterprise: state, church and companies. Moreover, the insertion of a *zone neutre* or *cordon sanitaire* between the European and African neighbourhoods in the early 1920s translated into spatial terms a colonial order based on the racial segregation that underscored Belgian colonial policy, especially in urban settings.[31]

In reality, however, this apparent colonial, binary structure was much more complex. Because of its very particular geographical location, situated on the crossroads of Central and Southern Africa, Lubumbashi was a cosmopolitan urban enclave, in terms of both the European and the African population.[32] Rather than being connected directly to the rest of Congo, let alone Kinshasa, the city was linked from 1910 onwards – the year of the city's foundation – to the outside world via a railroad coming up from Cape Town in South Africa via Bulawayo, a

31   On the aspect of racial segregation in the Belgian colonial context, a most useful study remains Georges Brausch, *Belgian Administration in the Congo* (Oxford: Institute of Race Relations, 1961).

32   For a late 1940s survey of Lubumbashi's population, see Ministry of Colonies, *Urbanisme au Congo belge* (Bruxelles: Ed. De Visscher, n.d. [1950]).

location in what is nowadays Zimbabwe.[33] Lubumbashi's urban culture thus was highly influenced by immigration from *l'Afrique australe*, while the large colonial enterprises operating in the city recruited their African labour from further afield. On top of this cosmopolitan aspect, social differences cut across Lubumbashi's urban society and structured relationships within as well as between the white and black urban communities.

This heterogeneity was translated both in spatial terms and in the built fabric of the urban landscape, demonstrating the extent to which Lubumbashi's urban space was a contested rather than a shared territory. The various groups within the white community of Lubumbashi seem to have been concentrated in specific areas. Italian, Greek and Portuguese traders, sometimes referred to as "second rate whites", for instance, were being located in those areas of the European neighbourhood nearest to the native town, as well as in a particular neighbourhood called Bakoa, outside its boundaries. The first urban plans of the cité Albert I, nowadays Commune Kamalondo, show a similar social stratification being projected onto an isotropic spatial grid, with "cilivized blacks" and "Hindus" being situated in those areas closest to the road that linked the "native town" to the European town.[34]

By erecting distinctive buildings on select locations within Lubumbashi's European town, the various European communities sought to affirm their identity clearly in the urban landscape. The palace of justice, the town hall, the gentlemen's club, the cathedral and the governor's residence, all situated along a main boulevard of the grid that crosses the Place Royale, the so-called Avenue de Tabora, marked the Belgian presence.[35] Italians built a consulate that by its building mass and stylistic treatment clearly stands out in the street. At almost exactly the same time, the Protestant missionaries erected a Methodist Church in neo-gothic style, while in 1956 the Greek community, who already possessed a clubhouse, inaugurated their orthodox church.

The important Jewish community in Lubumbashi had a synagogue built in 1929 (see Figure 10.6). Its particular location, situated on the extremity of the boulevard that symbolically represented the Belgian presence in the city – the synagogue "mirrors" the cathedral on the Avenue de Tabora – raises intriguing questions concerning the social position of the Jewish community vis-à-vis the Belgian colonial establishment,

---

33   It should be added that at that time another railroad also connected Lubumbashi to the city of Beira on the East coast of Africa.

34   An ongoing Ph.D. project by Sofie Boonen at the Department of Architecture and Urban Planning of Ghent University aims at establishing a detailed social geography of the city of Lubumbashi, using the Archives of the Land Registration as a main source.

35   The name *Avenue de Tabora* already points at the "Belgian symbolism" of this urban axis, as it refers to the famous victory of the Belgian colonial army over the Germans in East Africa in 1916, an episode that played an important role at the time in an effort of nation building, see Georges Delpierre, "Tabora 1916: de la symbolique d'une victoire", *Revue Belge d'Histoire Contemporaine*, 3–4 (2002): 351–381.

Figure 10.6 Synagogue, Lubumbashi, arch. Raymond Cloquet, 1929

Source: Photograph: Johan Lagae, 2005.

especially considering the important demographic shifts that occurred around the time of its construction.[36] From an architectural point of view, it is one of the most remarkable buildings in the city. Designed by the prominent colonial architect Raymond Cloquet, it forms one of the first examples of the introduction of modern brick architecture in the colony.[37] According to criteria common in the French practice of *patrimoine*, one could easily claim it to be a building of "national interest".

A once proud landmark indicating Jewish presence in the Congo, the synagogue has now stood vacant for decades, the Jewish community having declined drastically in numbers. Strikingly, the building has never been re-appropriated. Until this day it belongs to the small remaining Jewish community who secure its maintenance. This ongoing effort to hold on to the synagogue points at the value it still represents for the former and current members of Lubumbashi's Jewish community. It is the Jewish cemetery, however, that has become the privileged object of an explicit act of re-claiming a site as "heritage". A couple of years ago, the whole area was cleaned up and a website on the cemetery was posted on the Internet by Moïse Rahmani, who also authored a popularizing book on the history of the Jewish community in Congo, based largely on oral history.[38] By documenting the individual tombs through photographs and name indexes, the website invites people to add personal information (textual or visual) with relation to family members buried there. In such a way, the website turns this particular site at the outskirts of Lubumbashi into a virtual "lieu de mémoire". The example indicates to what extent a physical site can actually be re-appropriated from a distance by those who no longer live in the city. In that sense, it is not a heritage "shared" by former "colonizers" and "colonized", but a legacy that speaks to a very particular group that in fact surpasses the confines of Lubumbashi's former urban society. The website indeed turns the cemetery into a heritage site that addresses the worldwide Jewish diaspora.

In Conclusion

By discussing the theatre building and the Jewish cemetery, this chapter shows how investing meaning and adhering value to physical remnants of the colonial era is intrinsically linked to issues of identity and belonging. While it is important

---

36 Already in 1911 a *Congrégation Israélite* was founded in Lubumbashi, counting mainly Jews from South Africa among its members. In the late 1920s and 30s many of these would leave because of economic hardship, only to be replaced by the expanding immigration from mainly South-European Jews. This also induced a shift from a community with a primarily Ashkenazi background to one almost exclusively constituted of Sephardic Jews. See Jacqueline Benatar and Myriam Pimienta-Benatar, *De Rhodes à Elisabethville: l'odyssée d'une communauté Sephardic* (Paris: Ed. SIIAC, 2000).

37 On Cloquet's importance for colonial architecture, see Johan Lagae, "Raymond Cloquet", in *Dictionnaire de l'architecture en Belgique de 1830 à nos jours*, op. cit., 212–213.

38 www.sefarad.org/diaspora/congo/cimetiere, website compiled by Moïse Rahmani, author of *Shalom Bwana*, op. cit.

to understand the historical context in which such sites are rooted in order to assess their potential heritage value, this chapter argues that it is of equal importance to take seriously into consideration the "memory work" that is invested in them. Many other examples of Lubumbashi's built legacy could be given to make that claim. One can think for instance of what is probably the most important "lieu de mémoire" in Lubumbashi, the urban silhouette formed by the *terril* and chimney of the Gécamines, or the Union Minière du Haut Katanga as it was called in colonial times. The extensive research of the "Mémoires de Lubumbashi" group, as well as the recent photographic projects of Sammy Baloji, offer us fundamental insights in the way in which this particular urban site, once an icon of colonial propaganda, is still relevant to contemporary *Lushois*, albeit in very different ways.[39] This also became clear during the first heritage days in Lubumbashi in 2005, when the dramatic chances that the silhouette was undergoing due to recent re-exploitation of the former mining dump, became the subject of heated discussion.

By choosing sites that are not only lesser known, but that also act as "lieux de mémoire" for different communities that coexist but not necessarily interact, this chapter, however, aims first and foremost at breaking up some of the implicit assumptions embedded in the concept of "patrimoine partagé" or "shared heritage". It challenges in particular its underlying binary construction as a heritage shared by homogenized communities of former "colonizers" and "colonized". What is needed to overcome such assumptions, it is suggested here, is a return to that crucial and often unsettling question "Whose heritage?"

---

39   See, for instance, Donatien Dibwe Dia Mwembu and Bogumil Jewsiewicki, *Le travail hier et aujourd'hui. Mémoires de Lumbuashi* (Paris: L'Harmattan, 2004). For the work of Sammy Baloji, see Roger Pierre Turine, *Les arts du Congo d'hier à nos jours* (Bruxelles: La Renaissance du Livre, 2007), 140–145.

Epilogue

The Fragility of Memory and its Remedy Through Spatial Practices

Tali Hatuka

At the corner of Hayrkon Street they are destroying the house. The sea salt has eaten its iron tendons. Fifty two years old at its death. The workers begin, obviously, at the top, in the opposite direction of the house's creation. Already with the first hoe's hit, the upper room is exposed: it is the color of ideal flesh, the pink of the Twenties. Two or three prostitutes stand in the street in the morning, bored. The Mediterranean sea crouches in the distance, lazily licking itself. Just one important man, probably the contractor, carefully examines the grave being dug right in front of his eyes and foresees the future: here a square, multi-storied memorial will be built to the memory of the house.

(*Planning*, by Dan Pagis)

The poem *Planning*, by Dan Pagis, portrays the daily human process of erasing, constructing, modifying, forgetting and remembering places. Given that this process is a routine reality for so many people, why does it still have so much significance? Although there is no simple answer to this question, it is evident that a re-conceptualization is currently taking place – collective memory in social science and citizenship in planning practices – as two realms converge that together are creating a new engagement of citizens with memory in cities worldwide.

Associated with the debate on the limits of historical representation, the re-conceptualization of collective memory led to a body of knowledge that unsettled the established conventions of historical narration. Correlated with the work of the French sociologist Maurice Halbwachs, the contemporary definition of collective memory is conceived as a function of social power, and its expression varies with the social settings in which we find ourselves. For Halbwachs, studying memory is not a matter of reflecting on the properties of the subjective mind; rather, memory is a question of how minds work together in society, how their operations are structured by social arrangements (Olick, 2008). Halbwachs proposed that social groups – families, religious cults, political organizations and other communities – develop strategies to hold fast to their images of the past through places, monuments and rituals of commemoration (Halbwachs, 1992). Halbwachs's theory was rediscovered during the 1970s and 1980s with the expansion of collective memory studies, which became the debris of lost or oppressed identities, with scholars and citizens engaged in the excavations and genealogy of these identities.

This shift changed the role of collective memory, which became the raw material of social actions. As a result, collective memory became an elastic material, often remodelled, distorted, and hence made unreliable as a guide to the realities of the past. Memory became significant, not so much for its true representation but more as a social, political and cultural power and influence (Hutton, 2005). This led to the understanding of the role of memory in the making of political identities, as discussed in the work of Pierre Nora, for example, which addresses the making of the French national memory in the making of political identity (Nora, 1996), the work of Eric Hobsbawm and Terence Ranger (Hobsbawm and Ranger, 1983), which explored the political uses of tradition in the construction of collective identity, and the study by Benedict Anderson (Anderson, 1983) of the way "imagined communities" are constructed as public memories to give concrete affirmation to otherwise abstract ideals. The contribution of these scholars, and others, is critical to thought on the interlinked relations between memory and politics.

Parallel to the discourse on collective memory, and in association with it, we witness a shift in the conceptualization of citizenship. Generally, citizenship refers to a membership in a polity, contributing to the tension between inclusion and exclusion, between those deemed eligible for citizenship and those who are denied the right to become members. Three crucial features characterize the democratic political system: (1) the right to participate in the public sphere; (2) limitations in the power of government over the individual; and (3) a system based on the rule of law, not the arbitrary rule of rulers. With the turn of the twenty-first century, these features have been developed and enforced with governments focusing on the enhancement of civil participation and civil engagement as a tool, which reinforces democratic legitimacy and power. This approach to "the citizen" has significantly changed the planning discourse – from a passive subject whose projects are being planned for him to an active agent who participates in the development of the built environment.

The adoption of the participatory approach should be seen in the context of the failure of the utopian realization in the twentieth century, resulting in the disassociation of planning from the promise of utopia. This postmodern opposition to utopian projects championed everyday life and celebrates the civilian society (Chase et al., 1999; De Certeau, 1984; Lefebvre, 1984). Since the 1960s, planning has focused on the "here and now", objecting to all the concepts of utopia. Thus, planning adopted a dynamic framework influenced by a pragmatic approach to creating new visions. This has also affected the relationships between the professional, the citizen and the state. The citizen became a reference point, a player, an individual participating in the process of place-making, an approach that became part of the general agenda of inclusiveness and civic engagement enhanced by governments. This participatory, idealistic approach of the later 1960s has recently been replaced by a cruel realization that the target citizen of this approach is a member with legal status, whose support is needed to legitimize governance dominance by creating sanctioned space for participation (Miraftab, 2009: 43). Along with this, a rival approach in planning has developed focusing on the *idea of insurgency*. Coined by James Holston (Holston, 1998), the *insurgent* citizen refers to the individual who is challenging the hegemonic discourse and is able to initiate counter-hegemonic methods and tactics, by choice (or as last resort) in the process of place-making. Insurgent practices are often being hosted or encouraged by NGOs that gives the support, power and knowledge to actors in formal practices. Spatially, as argued by Ananya Roy, insurgency is closely associated with the idea of informality that is defined as a mode of production of space defined by the territorial logic of deregulation (Roy, 2009: 7–11).

Both discourses, citizenship and collective memory, are rooted in presentism – a counterpoint to the historicist idea of "progress", which dominated thinking about historical time during the modern age. The price paid for progress was the destruction of past ways of living and being in the world. Liberation meant active destruction of the past, which brought forgetting (Huyssen, 2003: 2). Presentism negates this by offering interpretations of the past that contribute to morally responsible, critical perspectives on the present age. From this perspective, history is no longer conceived as a grand narrative, or as a continuity that has informed the understanding of historical time in the modern age. The convergence of these discursive changes gave rise to numerous spatial practices of different actors (citizens, professionals and authorities) that transformed memory into an active, planned activity in cities worldwide.

How does collective memory become a tool in modifying space? With the growing significance of collective memory, places became the concrete sphere of negotiation over meanings. Moreover, citizens have the opportunity to negate or challenge the representation of future places and the way their symbols, memories and images will be conceived by professionals. Yet, in its essence, urban development often accentuates the power differences between groups because, by planning for the future, it challenges contemporary everyday life, and calls for transformation. This process entails change that relates both to the concrete construction of place and how it is integral to the cultural, national and political discourse of space. Thus, evolution of groups in the process of imagining place is often contested. These complex relationships between place, memory and spatial practices are infinitely repetitive and reversible, characterizing many of the processes of place-making.

Throughout the book, the authors offer various examples of relationships between collective memory and place, and comment on the use of spatial practices as a tool for different actors to establish meaning. Furthermore, fighting over meaning through spatial practices can take place through numerous methods, in particular by using practices of negotiation, reconstruction and performance.

Negotiation

The process of negotiation is inherent to most contemporary urban developments, as they are as much about forgetting as they are about remembering the past and present built environment. An example of negotiation over contested meaning is presented by Johan Lagae, in Chapter 10, when he discusses the recent debate over the built legacy of the former colonial city of Lubumbashi in the Democratic Republic of the Congo. The issue of the colonial legacy reminds us that Johan Lagae is not only talking about the conservation of physical remnants of the past but also about negotiating meanings embedded in these artefacts. Lagae calls for developing an approach that includes addressing both tangible aspects of the built environment (conservation, restoration) as well as the intangible aspects (memory, history). Another example in the book, which led to the active participation of rival groups, is the story of Yaad and Miaar in Galilee, Israel as presented by Tovi Fenster in Chapter 5. This urban development project aimed at erasing the memory of the Palestinian group in favour of an Israeli group of inhabitants, as part of the national project. Resisting the plan and bridging the process of collaborative planning, where both sides have learned to acknowledge the memory of the other, serve as a basis for change and future planning. In the process of negotiation over memory, often a mediator plays a significant role in bridging the gaps. In this particular case, and in most contested spatial developments, the professional or policy-maker often plays a crucial role in mediating between contested meanings, and integrating spatial production with political discourse.

Reconstruction

Spatial practices assist in reconstructing heritage, memories and, above all, a sense of (local/national) community. Yet, as noted by Richard Bauman, community stands for the kind of world that is, regrettably, not available to us – but that which we dearly wish to inhabit and to repossess. Community then becomes another name for paradise lost – but that which we dearly hope to return, and so we constantly seek roads that may bring us there (Bauman, 2001: 3). Reconstructing memory and a sense of community through the concrete environment is illustrated by Efrat Eizenberg, in Chapter 1, who offers an example of residents reconstructing elements of their past landscape in the community gardens of New York. In her study, Eizenberg shows how past landscapes that were part of the "environmental autobiography" of gardeners are treated by gardeners and in turn how these spaces create a positive connection to the gardeners' living environment. The gardens thus become spaces where individuals express their aesthetic and culture, typically not expressed within urban spaces.

Yet, parallel to the reconstruction in the scale of the community, we also witness processes of reconstructions at the scale of the state, which uses various methods to redefine its imagined sphere. An example of reconstruction at the scale of the state is presented by Damiana Gabriela Otoiu, in Chapter 8, who discusses the expropriations that affected the Jewish community after the installation of the Socialist regime and the reinstitution of urban property after the fall of that regime in 1989. During both time periods, the process of reconstruction of collective ideology was taking place using political and legal mechanisms that expropriate and then reinstitute property. In both cases, these mechanisms were used to reconstruct a collective national ideology and history. Here, property and the concrete space serve as a means to enhance the regime's legislative framework.

Performance

Speaking, listening and remembering are practices by which people transmit information. The capacity for speech and the associated capacities for learning and remembering might be thought of as the defining elements of human consciousness. The tradition of oral history served as a means to pass on knowledge and provide refuge for a group in some way marked as different from the rest of society. Acting out these stories publicly, through spatial practices, is a performing memory. As Elena Trubina reminds us, in Chapter 6, memory is no longer a transparent record of the past but rather a performative act. It is not neutral, morally or pragmatically, but place gives memory its contemporary meaning. Studying the case of Volgograd's built environment reveals an example of the Socialist tradition in modernist planning. Trubina illustrates how, for many, memory becomes the repository of "Soviet" memories, both by virtue of its urban structure and by the traumas of the Second World War. This state of affairs results in daily rituals through which works of memory have been performed in the city.

Performing collective memory could also be used as a tool of resistance as presented in the case of the Negev Bedouins. Safa Abu-Rabia, in Chapter 4, looks at how forcible spatial change (the war of 1948) has structured the Bedouin identity and how their attachment to their original lands serves as the basis of construction of their displaced identity, building feelings of conscious and physical alienation from the new space, which becomes an arena of resistance and protest. Abu-Rabia elaborates on the way Bedouins have attempted to build an exile identity with the intention of protesting and opposing their present situation, and expressing their continuous aspiration to return to their previous way of life and original land.

Whether using negotiation, reconstruction or performance to mark memory, the act of remembering is always in and of the present (while its reference is of the past and thus absent). Thus, inevitably, every act of memory carries with it a dimension of betrayal, forgetting and absence (Huyssen, 2003: 4). So, is this practice new? What characterizes the contemporary spatial processes of forgetting and remembering? First, the expansion of imagination's scope, with our horizons of time and space extended to include the local, national and international spheres, all defined elastically and undergoing constant changes. The elasticity of spheres allows creating new group coalitions, bypassing formal or existing borders. This is the beauty and the drawback of the imagination: it is flexible – a flexibility that generates a large umbrella under which more actors, organizations, citizens, communities, state authorities and international coalitions are included. Second, spatial processes of forgetting and remembering are used as tools for mobilization, and for the fight over resources and power. It is not the idea of "progress" or collective good in an abstract future that has driven these actions, but rather the temporal access to resources. Here, space plays a significant role, as it is visible and becomes a testimony, a manifestation of gained resources. Third, the role of memory and its spatial practices are in relation to historical trauma, when people try to come to terms with violence. In the twentieth century, we have witnessed episodes of genocide and mass destruction on a scale that has traumatized entire populations into a state of collective repression. Efforts to reckon with these horrifying memories have resulted in the creation of public memory spaces, including monuments, memorials, parks and collective rituals. Investigations of this phenomena have recalled Freud's thesis about the necessity of "working through" the trauma of repressed memory to uncover harsh and painful truths about crimes against humanity. In this context, urbanity plays a central role in the production of symbolic representations of the event and the place (Hatuka, 2009).

Finally, memory of all stripes remains a methodology involved in political. The complex and reciprocal interactions between space and memory not only create our urban landscape; they also transform our cities into a social property whose symbolism and iconography are constantly defined and recreated by its users. And yet with these characteristics of memory as an *active* spatial practice, it is important to recall the future, particularly when we try to envision alternatives to the contemporary situation. As Andreas Huyssen puts it:

> We need both past and future to articulate our political, social, and cultural dissatisfaction with the present state of the world. And while the hypertrophy of memory can lead to self-indulgence, melancholy fixations, and problematic privileging of the traumatic dimension of life with no exit in sight, memory discourses are absolutely essential to imagine the future and to regain a string temporal and spatial grounding of life. (Huyssen, 2003: 6)

In other words, buildings and monuments – designed by architects, planners and policy-makers in an endless process of production – define and change our landscape and establish a spatial array. This socio-spatial array forces us to adjust to particular social contexts, behavioural codes and political regulations. But, at the same time, this spatial array also provides us with a space in which to negotiate, oppose and resist. This particular dialectic of constraint and freedom is what makes urban spaces so crucial to memory practices, so strategic as a tool that allows people to negotiate their past – for, in the end, it is only through the imagination that we can envision a better future.

References

Anderson, Benedict. *Imagined communities: Reflections on the origin and spread of nationalism*. London: Verso, 1983.
Bauman, Zygmunt. *Community: Seeking safety in an insecure world*. Cambridge: Polity; Malden, MA: Blackwell, 2001.

Chase, John, Crawford, Margaret and Kaliski, John (eds). *Everyday urbanism.* New York: Monacelli Press, 1999.

De Certeau, Michel. *The practice of everyday life.* Berkeley: University of California Press, 1984.

Halbwachs, Maurice. *On collective memory.* Edited, translated, and with an introduction by Lewis A. Coser. Chicago: University of Chicago Press, 1992.

Hatuka, Tali. "Urban absence: Everyday practices versus trauma practices in Rabin Square, Tel Aviv". *Journal of Architecture and Planning Research,* 26/3 (2009): 198–212.

Hobsbawm, Eric and Ranger, Terence (eds). *The invention of tradition.* New York: Cambridge University Press, 1983.

Holston, James. "Spaces of insurgent citizenship". In *Making the invisible visible,* edited by Leonie Sandercock, 155–173. Berkeley: University of California Press, 1998.

Hutton, Patrick H. "Memory". In *New dictionary of the history of ideas,* 1418–1422. New York: Charles Scribner's Sons, 2005.

Huyssen, Andreas. *Present pasts: Urban palimpsests and the politics of memory.* Stanford: Stanford University Press, 2003.

Lefebvre, Henri. *Everyday life in the modern world.* London: Transaction Publishers, 1984.

Miraftab, Faranak. "Insurgent planning: Situating radical planning in the global south'". *Planning Theory,* 8/1 (2009): 32–50.

Nora, Pierre. *Realms of memory: Rethinking the French past* [*Lieux de mémoire*]. Edited and with a foreword by Lawrence D. Kritzman; translated by Arthur Goldhammer. New York: Columbia University Press, 1996.

Olick, Jeffrey K. "Collective memory". In *The international encyclopedia of the social sciences,* 2nd edition, edited by William A. Darity, Jr., 7–8. Detroit: Macmillan Reference USA, 2008.

Pagis, Dan. "Planning". In *Last poems,* 15. Israel: Hakibbutz Hameuchad Publishing House, 1987.

Roy, Ananya. "Strangely familiar: Planning and the worlds of insurgence and informality". *Planning Theory,* 8/1 (2009): 7–11.

Index

Page numbers in **bold** refer to figures